Breaking the Mold

Editors

David Millhorn

Stacey Patterson

Billy Stair

Design

LeJean Hardin

Photographic Coordination

Jason Richards

*Dedicated to the men and women of the
Department of Energy whose faith in us
made everything possible.*

From the Governor
GOVERNOR BILL HASLAM

Several years ago, I visited Oak Ridge National Laboratory for the first time in a long while. During the 20 minute drive from Knoxville, I reflected on the fact that Oak Ridge in general, and the Laboratory in particular, had always occupied a special place in Tennessee since their creation in the dark days of World War II. Growing up during the Cold War, my image of Oak Ridge had always been the "Secret City," a mysterious and slightly forbidding place that existed in many respects as an island to itself.

Like many recent visitors to ORNL, I remember being struck by the dramatic change in the Laboratory's appearance, so much so that for a moment I wondered if I was at the wrong place. Miles of rusting fences had been replaced by attractive stone markers that designated the Laboratory's major research programs. A modern research campus with new facilities and green lawns were where I remembered shabby buildings and ugly parking lots. The Secret City was open.

The pace and scale of ORNL's research program were even more impressive. In one of the new buildings, I saw two acres of supercomputers, capable of more than 1,000-trillion calculations per second, tackling problems for science and industry that only a decade ago were unimaginable. Up the hill, the simultaneous precision of 100,000 control points makes the Spallation Neutron Source an international center for the study of materials. It is no exaggeration to say that in less than a decade Tennessee has emerged as one of the world's foremost centers of scientific research.

This extraordinary success was not achieved by accident. The story is one of remarkable trust and cooperation among the Department of Energy, the University of Tennessee, Battelle, and the state of Tennessee. This partnership is even more remarkable given that it extended, in both Washington and Nashville, across administrations of both parties.

As Governor and Chair of the UT Board of Trustees, I am grateful for the invaluable synergy that UT-Battelle has brought to two of Tennessee's greatest assets. By combining the resources of a great laboratory and a great university, we have strengthened the research programs of both institutions while providing a critical contribution to the state's efforts to develop new companies and new jobs.

The most encouraging thing for Tennesseans is that despite the amazing progress made over the last decade at ORNL, I am convinced that UT-Battelle's most exciting days lie ahead. The foundation has been laid for a tremendous period of progress for the Department of Energy, for our flagship university, and for our state. I can only imagine what it will be like 10 years from now when I drive out to the Laboratory, reflecting on this unique partnership and knowing that we are solving some of the world's most important scientific challenges in Tennessee.

Table of Contents

Dr. Joe DiPietro

PRESIDENT, The University of Tennessee

At the University of Tennessee, we have always believed that we are defined by our aspirations. At no point in the University's 200-year history has our potential been fully reached or our desire to push the boundaries of knowledge satisfied.

The following pages are an account of the most recent milestone in our journey, our partnership with Battelle Memorial Institute to manage the Department of Energy's Oak Ridge National Laboratory, one of the world's foremost research institutions. Beginning with the decision in 1999 to compete for the Laboratory's management contract, we attempt to tell through our eyes a remarkable story about the people, the organizations, and, most important, the innovative ideas that together led to a resurgence of one of America's most valuable scientific assets.

The book's title, "Breaking the Mold," captures the philosophy that shapes UT-Battelle's approach to many of the most complex and challenging tasks one can imagine. Whether modernizing the Laboratory's infrastructure, building a world-class computational center, or helping the community renovate its high school, our tenure at the Laboratory has been marked by a readiness to undertake creative and unconventional strategies in support of the Department of Energy's research mission.

The progress we have made in just the last 12 years is astounding. Few people looking at the ORNL campus only a decade ago would have imagined that today it would be one of the most modern research institutions in the world, with an increasing number of capabilities among the very best in the international scientific community.

As proud as I am of this progress, I am even more excited about the prospects for the years ahead. In some respects, the Laboratory is just beginning to hit its stride, positioned better than at any time in its magnificent history to further the mission of the Department of Energy, strengthen the University of Tennessee's efforts to become a Top 25 public university, and provide one of the region's most valuable assets for economic growth.

Looking toward the horizon, the potential of Oak Ridge National Laboratory, like the University of Tennessee, will be limited only by our imagination.

A Historic Decision

> **"The chance to expand the University's partnership with ORNL was a risk we simply had to take."**
>
> —Dr. Joe Johnson, President, The University of Tennessee (1991-1999)

A year that proved to be among the most important in the distinguished histories of the University of Tennessee and Oak Ridge National Laboratory began 2,000 miles away in the Arizona desert. On January 4, 1999, the University of Tennessee defeated Florida State in the Fiesta Bowl, winning the National Championship for the first time since 1951. Sharing the jubilation with UT President Joe Johnson in his skybox were Vice President Al Gore, Former Senate Majority Leader Howard Baker, and Senator Fred Thompson. Also present was Bill Madia, Director of Pacific Northwest National Laboratory, a Department of Energy research facility managed by Battelle Memorial Institute of Columbus, Ohio.

The significance of the moment was not lost on the Tennessee delegation. With a high-profile victory over a major opponent, the University had established its place among the elite of college football. What the group that night also knew was that in two weeks

the University would be announcing a partnership with Battelle to engage in another competition of even greater national significance. This challenge involved something the University had never before attempted: develop a successful proposal to manage the Department of Energy's Oak Ridge National Laboratory, with the opportunity this time to establish for UT a place among the nation's elite academic research institutions.

The University's decision to pursue a multi-billion dollar management contract at one of the world's premier research institutions was not made lightly. In 1999 Lockheed Martin Energy Research had been the Laboratory's Management and Operations contractor for fifteen years. Lockheed Martin enjoyed the natural advantages of incumbency, including familiarity with the programs, facilities and personnel at ORNL. Perhaps even more important, fifteen years had produced a number of invaluable relationships between ORNL senior

Right page: *Ayres Hall, University of Tennessee. The University's relationship with Oak Ridge National Laboratory began during the Second World War.*

management and their counterparts at the Department of Energy. To say it would take a compelling proposal to overcome these advantages and convince the Department of Energy to replace an incumbent contractor was an understatement.

UT's ability as a southern land-grant university to compete against one of America's largest international corporations would be measured against the backdrop of a 50-year history of laboratory management dominated by a handful of prestigious academic institutions. A decision with such far-reaching implications would not be determined by political influence. The chance to manage America's largest energy research facility would require convincing the Department of Energy that UT deserved entry into a club that included the University of California at Berkeley, Stanford, Princeton, and the University of Chicago.

A 50-Year Collaboration

How the University determined it would compete for the ORNL management contract is a story with origins that go back to the Laboratory's earliest days in World War II. The Laboratory had enjoyed a special relationship with the University of Tennessee since 1943, when degree programs were established for dozens of researchers and engineers who had been transferred to Oak Ridge to work on the top secret Manhattan Project. Shortly after the Japanese surrender in 1945, Laboratory and UT officials established the first formal science

UT President Dr. Joe Johnson. His decision in 1998 to compete for the ORNL contract led to the formation of UT-Battelle.

The Decision

education partnership that enabled scientists to complete graduate studies at UT while working at what was then called Clinton Laboratories.

The Laboratory's relationship with UT continued after the war, encouraged by ORNL's Deputy Director for Research Eugene Wigner. Referring to research conducted at the national laboratories, the Nobel Prize recipient said, "As those wells begin to run dry, the situation becomes increasingly unhealthy." In 1948 UT Physics faculty William Pollard and Kenneth Hertel led the creation of the Oak Ridge Institute for Nuclear Studies, which later became Oak Ridge Associated Universities, a consortium of more than 80 universities that provides training and research opportunities in Oak Ridge for faculty and students.

In the early 1960s the partnership expanded to include faculty. Funded in part by the Ford Foundation, some 50 UT science faculty conducted research at the Laboratory and taught at the Knoxville campus. In 1967 UT President Andy Holt signed an agreement with the Laboratory and the National Institutes of Health to establish the UT-ORNL Graduate School of Biomedical Science staffed by Biology faculty holding joint appointments.

The partnership ascended to another plateau in 1984 with the creation of the Science Alliance, established by Governor Lamar Alexander as a UT Center of Excellence to promote research and educational collaboration between UT and ORNL. Funded entirely by state appropriations, the partnership included the Distinguished Scientist Program, the first effort of its kind in Tennessee designed to attract some of the world's premier researchers to joint positions at the University and the Laboratory.

The Rumors Prove True

Despite more than a half-century of shared faculty and students, University of Tennessee officials had never seriously entertained the idea of managing ORNL in the way that the University of California managed similar Department of Energy laboratories at Los Alamos, Livermore and Berkeley. The opportunity had been discussed briefly in 1984, but dismissed. The issue surfaced again during the spring of 1998 with recurring rumors in both Washington and Oak Ridge hinting that the Department of Energy would not extend Lockheed Martin's management contract at ORNL. If the rumors proved true and the contract would be competed, the implications for both ORNL and the University were enormous.

In Oak Ridge, a sense of anxiety prevailed in a community with an economy closely tied to the Laboratory's long-term viability. Years of deferred maintenance and little capital investment had produced a campus

that was outdated and expensive to operate. Meanwhile, the Congress had funded new facilities and programs at other DOE laboratories in Brookhaven, Pacific Northwest, and Argonne. A report commissioned by the Congress in the mid-1990s suggested ominously that in the event budgetary constraints forced the Department of Energy to close a laboratory, ORNL would be a likely target.

Twenty miles away in Knoxville, University officials shared a similar anxiety. More than many on the UT campus realized, the growth potential of the University's research portfolio, particularly in the areas of materials and high-performance computing, was closely linked to the vitality of comparable programs at ORNL. Unless reversed, the gradual decline of ORNL's relative standing in the DOE

UT President Dr. Andy Holt. The creation of the UT-ORNL Graduate School of Biomedical Science in 1967 laid the foundation for the UT-Battelle partnership three decades later.

5

laboratory system threatened short-term implications for UT's academic programs and long-term damage to the University's reputation as a research institution.

A host of other rumors—many never verified—compounded concerns about a potential change of contractors at ORNL. Among the most alarming had the University of Texas making plans for the contract competition. The idea that the "other UT" might assume control of Tennessee's largest research institution in the shadow of the Knoxville campus would have been an unimaginable blow to the psyche of the entire region. Regardless of their validity, these and other rumors contributed to a decision by UT officials to explore more seriously than ever before if and how the University should seek a management role at Oak Ridge National Laboratory.

Shortly after spring graduation in May 1998, UT President Joe Johnson assigned the task to Homer Fisher, the University's Senior Vice President closely involved in the Oak Ridge community. More than anyone at the University, Fisher knew the personalities and understood the unique culture of both the Laboratory and its landlords at the Department of Energy. In early August the Department of Energy officially announced plans to conduct a competition for the ORNL contract. After discreet meetings and conversations with a network of interested parties, Fisher convened a meeting on August 18 of senior UT officials. The discussion was intense. One official argued that because of budgetary pressures UT could not afford to bid on the ORNL contract. Given the potential consequences, Fisher countered that UT could not afford not to bid. Making one of the most important decisions in

the University's 200-year history, President Johnson ended the debate. He instructed Fisher to lead the development of a proposal to the Department of Energy to manage Oak Ridge National Laboratory, taking the University down a road never before traveled.

Choosing a Partner

UT's decision to compete for the ORNL management contract was followed immediately by another decision. While the University could offer the Department of Energy a significant contribution to ORNL's scientific agenda, as a practical matter UT had little experience managing the unique operational challenges of a national laboratory that contained a nuclear reactor and a host of radiological facilities, an extensive inventory of classified assets, and security policies unfamiliar to an open academic campus. To fill the need for operational

management experience at a laboratory with nearly 4,000 employees and hundreds of buildings, UT began the search for a corporate partner.

Finding the right match was no simple task. Credibility required a partner with experience managing institutions comparable in size and complexity to ORNL. A laboratory budget tied closely to federal appropriations demanded a partner familiar with the halls of Congress. Ideally, the partner would be big enough to be a player on the national stage but small enough that ORNL would be a priority in their corporate portfolio. Last, and perhaps most important, was a chemistry that indicated an understanding and appreciation for the culture of UT and ORNL.

The late summer and fall of 1998 were filled with dozens of discussions, presentations and negotiations with parties interested in teaming

Above: *Battelle Memorial Institute, Columbus, Ohio. Relatively unknown in Tennessee in 1999, Battelle enjoyed a national reputation for laboratory management and technology transfer.*

with UT on the ORNL contract. One priority that emerged from these discussions was UT's desire to establish a closer partnership at ORNL with a consortium of premier universities whose collective assets could enrich both the Laboratory's research agenda and educational outreach. The effort was led by Oak Ridge Associated Universities, whose new president Ron Townsend ultimately negotiated formal agreements with Duke, Florida State, Georgia Tech, North Carolina State, Virginia and Virginia Tech to become members of the "Core Universities" on the UT team.

Taking place in parallel with the university discussions, UT explored potential teaming relationships with some two dozen potential partners, including Lockheed Martin. In an atmosphere where discretion was important, some meetings took place in the University's board room while others were conducted in airport

hotels. On October 26, two days after a 35-18 victory over arch-rival Alabama, UT entertained Bill Madia, Executive Vice President for Battelle and Director of DOE's Pacific Northwest National Laboratory in Richland, Washington. The nation's largest not-for-profit research institution, Battelle had a successful history of managing Department of Energy laboratories in Upton, New York and Richland, Washington. Battelle proposed a teaming arrangement similar to one used between Stony Brook University and Brookhaven National Laboratory in which Battelle handled the majority of operational responsibilities and the university focused on the laboratory's research agenda.

On December 4, UT sent a letter to Battelle officially proposing a teaming arrangement that would create a 50-50 partnership. The following weeks, including the holidays, were filled with quiet negotiations that

methodically sorted through contract details such as liability, pension plans, and, most important, who would lead actual development of the contract proposal. On January 8, Duke and North Carolina State formally joined the team, becoming the first two Core Universities. On January 20, two weeks after winning the football national championship, the University issued a press release announcing the formation of a new entity, UT-Battelle, LLC, created for the sole purpose of competing for the management contract of Oak Ridge National Laboratory.

For the University and for the Laboratory, a new era was underway.

A Groundbreaking Proposal

> **"**This creative proposal is one of the best answers I have ever heard to those who say it is impossible for the federal government, state government and local government to work together.**"**
>
> —Governor Don Sundquist

Confronted with the task of unseating a fifteen-year incumbent, the UT-Battelle proposal team believed the Department of Energy could be convinced to change contractors only if presented with a number of genuinely innovative and transformational ideas for managing ORNL. If UT-Battelle's management proposal promised only more of the same with a different set of faces, DOE would be less inclined—regardless of current contractor performance—to incur the risk that unavoidably accompanies the changing of horses in mid-stream. Compounding the challenge was the highly-structured nature of DOE's Request for Proposals, a complex document that required detailed responses to dozens of items in the management categories of research, laboratory operations, and ORNL's civic and industrial outreach activities.

The UT-Battelle proposal team began work immediately following the January announcement of the intent to bid.

Right page: Governor Don Sundquist. Open support from the Governor and the legislative leadership was an unexpected addition to UT-Battelle's management proposal for Oak Ridge National Laboratory.

From the first, the team's organization reflected the working relationship envisioned by the two parties. UT's efforts were led by Homer Fisher, who coordinated input from the science leadership of the Knoxville campus, including the Vice Chancellor for Research, the Dean of Engineering, and the department heads for Physics, Biology, Computer Science and Civil Engineering. Battelle's contribution was led by Bill Madia, who made available a number of senior staff from Pacific Northwest National Laboratory to address operational issues such as protocol for radiological facilities, security of classified assets, and a host of other items unique to the management of a large government research institution. Meanwhile, throughout the spring Fisher and Madia also handled dozens of meetings and conference calls with companies and institutions interested in joining the UT-Battelle team in a more limited role as subcontractors.

The UT-Battelle proposal team determined that the chances of winning the competition depended upon providing the Department of Energy a compelling management plan divided into two distinct categories.

The first category would contain UT-Battelle's response to specific issues and expectations contained in DOE's Request for Proposals. DOE would evaluate and contrast all bidders on the basis of their written and oral responses to these issues. The second, and in some respects more challenging category of the plan, consisted of innovative ideas that UT-Battelle would offer in the areas of Science and Technology, Laboratory Operations, and Community Outreach. The UT-Battelle team increasingly believed that developing a number of compelling and credible new management approaches at ORNL was critical to success against an incumbent contractor.

Meeting Expectations

With one significant exception, DOE's management expectations for ORNL reflected the agency's goals and challenges throughout the national laboratory system. DOE wanted potential contractors to align ORNL's research portfolio more closely with the Department's long-term mission, improve safety, reduce costs, enhance the economic impact of scientific research, and establish more

collaborative partnerships with higher education and industry.

In 1999 the most unique issue in the ORNL Request for Proposal was the construction of the Spallation Neutron Source, a $1.4 billion proton accelerator intended to reclaim from the Japanese American leadership in advanced materials research. In Oak Ridge, building and operating the world's largest science facility would provide a scientific anchor and remove doubts about the Laboratory's long-term viability. In Washington, delivering the SNS on time and on budget, particularly after the recent collapse of the Superconducting Supercollider project in Texas, would go far in restoring congressional confidence in the department's ability to manage the construction of mega science projects. Put simply, DOE viewed the SNS as a project that could not be allowed to fail. After years of delay, the project was scheduled to break ground in late 1999, just weeks before the contract transition. Changing management of the project at this crucial juncture would come with enormous risk. Not apparent to most observers, the need to convince DOE of management stability for the SNS project was among the single greatest obstacles for the UT-Battelle team's chances of unseating an incumbent contractor.

Two Bold Commitments

UT-Battelle responded to DOE's core management expectations with a number of bold commitments. Among the most notable was a pledge to reduce overhead costs by 20 percent within three years and dedicate the savings to growing the R&D portfolio. ORNL's high rate of accidents would be lowered through implementation of "integrated line management," a strategy that sought to define

more clearly management's roles and accountability for employee performance, including safety. Coordinated by Oak Ridge Associated Universities, academic collaboration would be enhanced through the establishment of a group of six "Core Universities"—Duke, Florida State, Georgia Tech, North Carolina State, Virginia, and Virginia Tech—that would help shape the Laboratory's research agenda and establish direct links to student and faculty research opportunities at ORNL. The proposals were solid, but hardly dramatic.

While the UT-Battelle team felt that as a practical matter it would be difficult to develop a distinctly superior proposal based solely upon conventional management issues, they had no such reservations about a series of other proposals that recommended a fundamental realignment of ORNL's relationship with UT and with the state of Tennessee. Since the earliest days of the Manhattan Project, the Laboratory had operated both literally and figuratively "behind the fence." Not surprisingly, six decades of working in such a closed environment gradually shaped the culture of the ORNL staff. An institution that operated behind guards and gates was not instinctively open to sharing the Laboratory's talents and technologies. UT-Battelle sought to reverse 60 years of operational practice with an ambitious proposition. While delivering the research mission for the Department of Energy customer, the Laboratory would leverage its magnificent research facilities by

Right: *Entrance gates, the Laboratory's first impression near the old visitors center. UT-Battelle believed the removal of miles of rusted fences and signs was critical to changing the Laboratory's image and culture.*

UT Senior Vice President Homer Fisher. His personal skills played a key role in UT's ultimate partnership with Battelle and the subsequent development of a successful contract proposal.

making these unique assets available as never before to large numbers of researchers from universities, industry and other national laboratories. In addition, UT-Battelle would make a comparable commitment to removing the barriers and providing the resources needed to commercialize the cutting edge technologies developed at the Laboratory.

The proposal's potential for the University of Tennessee and for the region's economy was enormous. Students and faculty would be granted access to world class facilities and equipment at ORNL that the state could never hope to duplicate at the University. Plans to expand the number of joint faculty meant that UT could recruit from a broader and more prestigious pool of candidates drawn to facilities such as the Spallation Neutron Source. Similarly, a larger and academically richer number of outstanding graduate students in the scientific disciplines would be attracted to the possibility of internships and post doc positions at one of the world's most prestigious research laboratories.

UT-Battelle's ambitious plans were also designed to appeal to state officials by positioning ORNL as an engine for economic growth in the East Tennessee region. The proposal contained a commitment to generate ten new start-up companies annually from Laboratory technologies. UT-Battelle proposed to allocate a substantial portion of its contract fee to funding a new technology transfer program designed to assist Laboratory researchers in dealing with the legal, financial, marketing, and other obstacles that often impede the creation of new companies. Dwarfing in scale any similar program ever contemplated at ORNL, UT-Battelle envisioned over the next decade the creation of dozens of new companies and hundreds of new jobs to complement the state's economic development strategy.

As far-reaching as these closely-held ideas were, in the spring of 1999 they represented only a portion of UT-Battelle's plan to revamp and expand ORNL's roles for educational and economic outreach. A second and even more ambitious proposal involved

securing a commitment from the state to undertake a major capital investment in the Laboratory's deteriorating infrastructure. In 1999, the idea of a state government funding large research facilities at a federal laboratory was unheard of. To many, the notion that a relatively poor state such as Tennessee would allocate precious resources to ORNL seemed completely unrealistic.

UT-Battelle's ability to conceive and develop a plan that called for the state of Tennessee to build three new facilities at Oak Ridge National Laboratory reflected, more than any other single initiative, the creativity

and boldness of the new partnership. After conversations with UT officials, UT President Joe Johnson in May 1999 received a letter co-signed by Governor Don Sundquist, Senate Speaker John Wilder, and House Speaker Jimmy Naifeh pledging $18 million in state funds (later increased to $26 million) for the construction of three facilities on the ORNL campus. Significantly, the letter stated that, "These commitments are exclusive to UT-Battelle contingent upon the contract award." The ability to secure an unprecedented level of state financial support—exclusive to UT-Battelle—was a testament to the affinity enjoyed by the University of Tennessee among state government leaders and an early indication of the changes that UT would help bring to the Laboratory's research agenda.

The significance of the state's commitment to build three buildings on the ORNL campus extended far beyond the research programs contained in the facilities. Implicit in the agreement was a strategy to leverage ORNL's resources and capabilities in a way that had never been contemplated. A federal research facility, with one of America's largest concentrations of intellectual talent and technological innovation, would become a full partner with the state of Tennessee and the state's flagship university. For UT, the vision would bring invaluable support to the University's goal of becoming a top 25 public institution by creating dozens of new joint faculty appointments with expanded access to some of the world's most sophisticated research facilities. For the state, the partnership's commitment to technology transfer and commercialization represented an extraordinary addition to Tennessee's economic development strategy.

For ORNL, the two partners were about to change the game.

"UT-Battelle and I always shared a simple philosophy. Be bold and say, 'Why not?'"

—*Congressman Zach Wamp*

Upper right: *Congressmen Bart Gordon and Zach Wamp. Gordon's efforts in 1999 saved the Spallation Neutron Source project from termination. Wamp's support was an important factor in the success of UT-Battelle's agenda.*

Previous page: *The "Winter Palace." Home to UT-Battelle's new leadership team during the 90-day contract transition, the Manhattan Project era Quonset huts came to be a symbol of the Laboratory's need for modernization.*

In October 1999, approximately 18 months since the initial discussions about the ORNL contract competition, University officials eagerly awaited the phone call from the Department of Energy that would inform them whether their efforts had been successful. Those closest to the process were cautiously optimistic that intense preparation for the written and oral presentations would pay off. Others were worried whether DOE could be convinced to replace an established contractor. Still riding the emotional high three days after the Vols' nationally televised 21-7 win over Alabama in Tuscaloosa, the anxiety turned to elation on October 20 when the call came from DOE headquarters announcing that UT-Battelle had been awarded a five-year, $2.5 billion management contract for Oak Ridge National Laboratory. For the University's research program, the victory was as significant as the football national championship earlier that year.

The time allotted for celebration was brief. In roughly 60 days—a period that included the Thanksgiving and Christmas holidays—DOE would begin the official transition of management operations from Lockheed-Martin to UT-Battelle. The 90-day transition would involve the replacement of 15 of the top 18 management positions at the Laboratory. Despite access to extensive operational and financial data, only a few members of the new Leadership Team had spent significant time on the ORNL campus. Becoming familiar in only twelve weeks with a complex that contained hundreds of buildings and nearly 4,000 staff was a daunting task.

The first public event involving UT-Battelle occurred on December 19 prior to the actual transition when DOE Secretary Bill Richardson joined elected officials in a ground-breaking ceremony for the $1.4 billion Spallation Neutron Source. Attended by Vice

President Al Gore, Tennessee Governor Don Sundquist and ORNL alumnus and Nobel Prize recipient Clifford Shull, the event was a nightmare for the Secret Service, which had to organize security for an outdoor event in freezing temperatures at the top of a wooded ridge accessible only on a small one-lane dirt road. Once the guests were assembled, the ceremony had a surreal atmosphere. An event intended to symbolize and celebrate a resurgent future for ORNL's research program was conducted by senior Laboratory managers who would soon be losing their jobs.

Management transitions are always accompanied by a degree of awkwardness, and, on occasion, by bitterness, the unavoidable reaction to losing a hard-fought, high stakes competition. The UT-Battelle transition team found it hard to believe that some of these lingering emotions were not present on January 3rd when they

moved into their temporary "offices" on the ORNL campus. Located in a rusted Quonset hut used after World War II to examine the impact of radiation on humans, the transition staff had been assigned what was possibly the worst facility in the Laboratory complex. With sporadic heating and squirrels and possum nesting in the attic, the building actually had a positive unifying effect on the new management team, who christened the facility the "Winter Palace" after the home of the Russian czars in St. Petersburg.

An Immediate Crisis

The most serious problem confronting the UT-Battelle transition in January 2000 had little to do with assuming operational control of the Laboratory. The celebration and speeches at the recent SNS groundbreaking ceremony had ignored the distinct possibility that the massive facility might never be more than a huge

pile of dirt. The previous May, House Science Committee Chairman Jim Sensenbrenner of Wisconsin had come within a single vote of killing the nation's premier science project. The project was given a nine-month extension through a last-minute compromise negotiated during the lunch recess by Tennessee Congressman Bart Gordon. Sensenbrenner essentially agreed to put the SNS on life-support until February 1, 2000, when the project would be required to meet seven stipulations adopted by the committee. Failure to comply with the committee's expectations would likely result in the project's termination. Among the stipulations was confirmation that the state of Tennessee would not impose the current sales tax on transactions associated with the SNS construction. Sensenbrenner granted the extension knowing that compliance would require passage of a new statute exempting the SNS project. He probably also was

aware that the Tennessee legislature had adjourned for the year and would not reconvene until mid-January 2000.

Faced with an impending crisis during the first week of the transition, the joint response by UT and Battelle was a dramatic signal that the new team would take a bolder and distinctly different approach to dealing with the Laboratory's operational challenges. Fearful that a political solution was out of the question, Lockheed-Martin had been exploring a legal challenge to the congressional demand that the sales tax on SNS construction be removed. Battelle, headquartered in Columbus, Ohio, was relatively

unknown in Nashville. The University of Tennessee, in contrast, enjoyed enormous affinity among the leaders of the Administration and the Legislature. Ignoring the high risk of failure, UT sought to do what many thought impossible—meet the February 1 congressional deadline by introducing and passing a $28 million piece of legislation in less than three weeks.

Securing the bi-partisan support of the Governor and the two Speakers— both UT graduates—the University threw its weight behind legislation that would provide a sales tax exemption on construction activities associated with the Spallation Neutron Source.

Suspending the House and Senate rules at every opportunity, the bill's sponsors moved the legislation at what the media called "warp speed." The final vote came with House approval on the morning of January 28. The bill was engrossed that afternoon, signed by the Governor the next day, and transmitted overnight to Congressman Sensenbrenner's office in Washington, beating the deadline by 24 hours.

The significance of the SNS legislation was two-fold. In Washington, the project gained new life, gradually winning the confidence of congressional appropriators and eventually restoring the Department of Energy's reputation

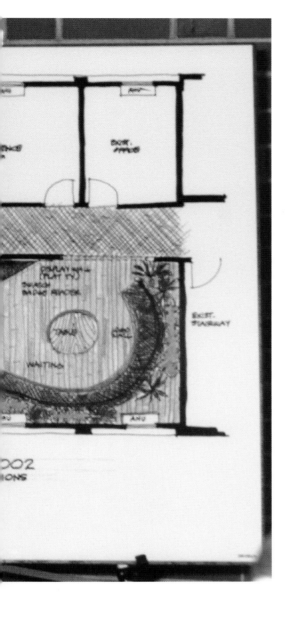

as an agency capable of delivering mega science projects on time and budget. Equally important, UT-Battelle's ability to pull the SNS out of the political fire seemed to have a similar positive impact in the Oak Ridge community and at the Laboratory. In a way that was never fully credible with receptions and press releases, the message came through that genuine change might be on the horizon.

Changing the Mindset

Back at the ORNL campus, UT-Battelle viewed the Winter Palace as the visual symbol of the Laboratory's need for transformation. The new

management team arrived aware that ORNL's infrastructure had suffered from years of deferred maintenance and limited investments in new buildings and renovations. They were shocked, however, at the extent to which many of the laboratories, offices and other facilities had been allowed to deteriorate. Dozens of buildings dated from the 1950s and 1960s, and looked the part. Many were cramped and inefficient, with heating, ventilation and plumbing systems that worked sporadically at enormous cost. UT-Battelle discovered that for some of the Laboratory's research programs, the facilities were so bad potential employees were no longer brought to ORNL for a site visit.

Apart from these infrastructure problems, the campus was simply visually unattractive, looking more like a run-down factory than a modern research institution. The central campus was surrounded by miles of forbidding fences and turnstiles. Roads and buildings were blighted with hundreds of rusted signs of all sizes and colors whose purpose had been long forgotten. Landscaping was scarce. Weeds had overgrown creek banks and empty lots. The sight and smell of restrooms, even in ORNL's cafeteria, suggested employees over time had become resigned to a substandard quality for the majority of the Laboratory's facilities, perhaps believing things had been so bad for so long that they would never change.

The impact of ORNL's aging infrastructure on staff morale, as well as on the increasing difficulty in recruiting world-class talent, was immediately obvious to the new management team. The problem's scale was such that it could not be solved with a single new building or renovation. Reversing decades of decline would require a

dramatic initiative, large enough in scope to be credible and bold enough to convince long-time employees that genuine change was possible. The first message of pending change was sent with a press conference held on a rainy winter afternoon in the Laboratory's Visitors Center, a dingy and outdated building that symbolized the image of ORNL's infrastructure. Decorated with cheap furniture and posters and located by a rusted fence cluttered with signs, the cramped little facility was the Laboratory's first impression for thousands of visitors each year.

To the surprise of many staff who had walked by the building each day for decades, the new Lab officials declared they would tear down and replace the ORNL Visitors Center. A nearby building was renovated to provide visitors a spacious waiting room with modern furniture, Laboratory artifacts, and walls tastefully decorated with large murals of the region's mountain scenery. The opening of a modern new Visitors Center, followed by the demolition of the old eyesore, delivered a powerful statement that UT-Battelle would seek a different path for the Laboratory's operation.

Lab officials recognized that replacing the Visitors Center, while important, was only a symbolic gesture. They understood that revitalizing the Laboratory's infrastructure would be possible only with a large number of new facilities. They also understood that transforming ORNL's industrial image would require something more thoughtful than the ad hoc approach of buildings, asphalt parking lots and fences that characterized the current campus. For help they turned to the University of Tennessee College of Architecture. Professor Jon Coddington was hired to develop a master plan for the ORNL campus, the first such plan that attempted to provide a coherent approach to the design, location and long-term planning of buildings, roads and parking at the Laboratory.

Coddington's recommendations were more ambitious than many expected. They included removal of an enormous thousand-car parking lot, with parking moved outward to lots on the campus perimeter. The former parking lot would become the center of a new East Campus, designed to resemble a college with a "quad" of grass surrounded by new buildings constructed with complimentary architecture. Several blocks west a similar motif would be used to align ORNL's biology facilities into a West Campus. Miles of fences and hundreds of signs would be removed. Some roads for the first time would be restricted to pedestrians and bicycles. Much of the new campus would be accented by flowers, trees and waterfalls.

While portions of Coddington's master plan were modified over time, the University had contributed to a fundamental and lasting shift in ORNL's operational mindset. Going forward, UT-Battelle's adoption of a new approach to architecture and landscaping meant the Laboratory was beginning to think of itself less as a Cold War facility and more as a modern research institution. For ORNL, the implications of this change, on the research agenda as well as the infrastructure, would be profound.

Right: *"New" visitors center. Remodeled from an existing building and eventually replaced in 2005 by a new facility, the renovated visitors center in 2000 was used by UT-Battelle as a visible symbol that change at ORNL was possible.*

Year One

"We believed if we could bring about a fundamental change in the Laboratory's culture, there was no limit to ORNL's potential."

—Bill Madia, Laboratory Director, 2000–2003

While the new UT-Battelle management team arrived at ORNL with a number of long-term goals for the Laboratory's operational and research agendas, it was clearly understood that the first order of business would be responding to performance expectations established by the Department of Energy in the recent contract competition. As the owner of the facilities and the funding source for the large majority of ORNL's research programs, DOE is frequently referred to as the Laboratory's "landlord." UT-Battelle preferred to view DOE as a "customer," a term intentionally selected to instill among management and staff the belief that the Laboratory each day was providing an array of services requiring the same level of quality and customer relations as a major private corporation. Implicit in this philosophy was the idea that the most productive relationship with DOE was one based on trust that would be less adversarial and more collaborative than in the past.

Building trust with the DOE customer would require immediate attention to the Department's primary operational

concerns at ORNL. Again, UT-Battelle had its own nomenclature. The process of responding to specific DOE concerns and directives was called "answering the mail," a term whose connotation included an expectation to address such issues quickly and fully on a priority basis. In the spring of 2000, UT-Battelle's initial agenda by necessity was shaped around the need to answer the mail on two issues of particular importance to the Department of Energy. Not surprisingly, addressing each challenge would require a fundamental change in the Laboratory's culture.

Improving Safety

Few performance issues are of greater concern to DOE officials than the safety of the public and laboratory staff. In 2000, ORNL's safety record had consistently been among the lowest in the DOE laboratory system. Accidents recorded by employees were measured in two ways. Relatively minor accidents such as cuts and sprains fell into one category. The Laboratory also kept statistics for more serious safety incidents that resulted in days missed

Previous page: *A high rate of accidents among ORNL staff was one of UT-Battelle's greatest challenges in 2000.*

Right page: *Demolition of Building 1000. Erected as a "temporary" office building in the 1940s, the facility was typical of a Laboratory infrastructure that was old, ugly and expensive to operate.*

from work. ORNL's safety record was historically weak in both categories, a pattern of concern to DOE and one the agency insisted be improved.

ORNL's safety performance prior to 2000 resulted largely from two factors. The Laboratory's daily activities include a wide variety of activities that range in complexity from roadside mowing to the handling of highly radioactive materials. Despite the risks posed by many of these tasks, a general indifference to safety, like a number of other operational issues UT-Battelle encountered, was the product of a culture that had evolved over the decades among Laboratory staff at all levels. Whether it was the reluctance of craft workers to use safety glasses, the ignoring of safety work plans by lab technicians, or the refusal to obey speed limits on the Laboratory's roads, in each instance the behavior reflected

a basic lack of appreciation for safety's importance. UT-Battelle's management team believed that such behavior is not changed overnight, and certainly could not be reversed with strategies that rely solely on posters and short-lived safety campaigns.

As is often the case, a single incident served as the pretext to undertake a major effort to reshape the safety culture at ORNL. Breaching multiple safety rules, a lab technician seriously mangled his hand with a saw. The response to the accident represented a fundamental change, both in safety expectations and in the way the Laboratory implemented incentives for improved safety performance based on the notion of accountability. In effect, safety became a performance expectation for each manager, with annual compensation based in part on the safety record—including

speeding—of supervised employees. The Lab Director produced a safety video and required each manager to watch the video with staff. Safety work plans were developed and enforced. The Laboratory's director of Facilities and Operations met individually with each new craft employee to emphasize safety expectations. Counter to past practice, a small number of staff with repeated or especially egregious safety violations were terminated. On the positive side, craft workers who went a year without a recorded safety incident received small bonuses and became eligible for a drawing to win a new truck.

Gradually, the Laboratory began to experience a sustained shift in safety statistics. From the bottom, then to the middle, ORNL eventually reached the plateau among the labs with the best safety performance in the DOE system. By 2011, UT-Battelle could announce

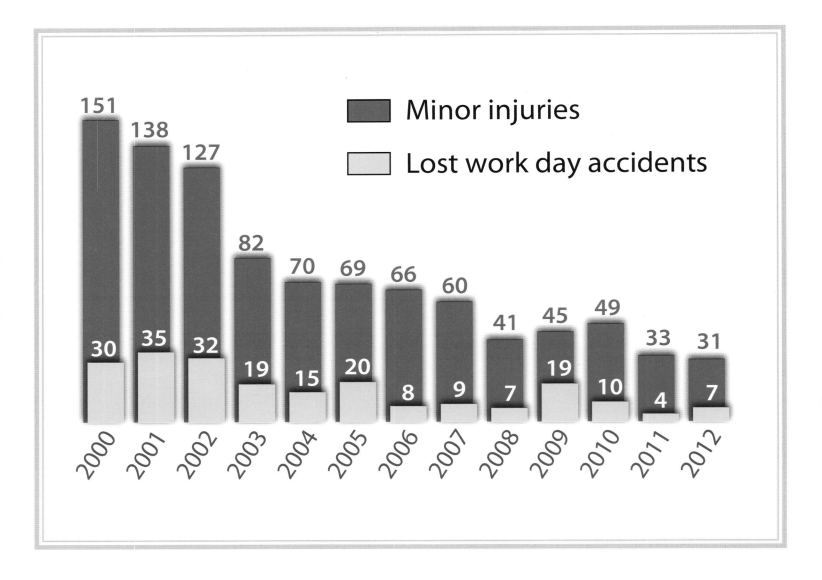

that lab employees had worked 7 million consecutive hours without a serious injury, a milestone unimaginable only a decade earlier.

Reducing Costs

One of the most ambitious commitments contained in UT-Battelle's bid proposal was a pledge to reduce overhead costs by 20 percent in the first three years. During the first all-hands meeting in April 2000 the goal was met with open skepticism among some Laboratory staff. Many long time employees had heard such promises before and viewed the pledge as the kind of rhetoric common in contract competitions but quickly abandoned.

The new management team had a different perspective. Faced with an aging infrastructure and the likelihood of relatively flat budgets for the foreseeable future, the only realistic way to provide more resources for research staff required a substantial and sustained reduction in the cost of operating the Laboratory.

UT-Battelle determined that any chance of realizing meaningful cost reductions depended upon the ability to implement major improvements in efficiency and lower the Laboratory's utility and equipment expenses. Although it was less obvious to existing staff, UT-Battelle inherited a management and operational

structure that over the previous two decades had accumulated numerous low-value programs and expanded numbers of management personnel. The Laboratory was divided among researchers funded directly by grants and DOE appropriations, and the remaining staff comprised of managers, secretaries, laborers and others who provided support functions for the research programs and who were funded from research allocations. Ideally, UT-Battelle believed a 60-40 ratio of research to support staff should be the Laboratory's goal. In 2000, the Laboratory's ratio was reversed, with three support staff for every two researchers. The inevitable result of this ratio was a high cost for ORNL research.

If the Laboratory wished to provide greater investments needed to grow the research programs, the effort would require a significant reduction in the number of support staff. The first twelve months of the UT-Battelle contract witnessed an extensive effort to identify low-value work and generate cost-saving recommendations. In the spring of 2001, UT-Battelle announced a voluntary staff reduction, followed by a smaller involuntary reduction that together totaled more than 200 staff. The staff reductions were accompanied by a strategy aimed at holding overhead costs to one percent annual increases while growing the Laboratory's research portfolio by two percent above inflation each year. With minor adjustments, UT-Battelle held to the plan for a decade. By 2012, the incremental changes had produced a reversal of the 2000 research-to-support staff ratio. The result was millions of dollars available to upgrade research programs and invest in new facilities.

UT-Battelle's effort to reduce overhead costs also included a sweeping plan to reduce the size of the Laboratory's footprint. In 2000 ORNL facilities represented more than 4.4 million square feet of space that required heating, cooling, cleaning and maintenance. The high cost of providing these services was related not only to an excessive amount space but also to the fact that 100 percent of the buildings at ORNL were more than 10 years old, and nearly one-half of the buildings were more than 40 years old. Approximately 25 of these buildings, most of which were built in the 1950s, accounted for 75 percent of the Laboratory's $60 million in annual power usage. The new management team determined that substantial savings would be possible only if the volume of space could be reduced and, if possible, replaced by newer and more energy efficient facilities.

Contained in UT-Battelle's cost reduction strategy was a goal of dramatically reducing the Laboratory's footprint of high-cost space that was of marginal value. Implicit in the strategy was an ambitious plan for renovation or demolition of old buildings and the construction of new ones. Working with the University of Tennessee's College of Architecture, plans were developed that envisioned a dramatic new campus with facilities that were planned for the first time with thought given to their operational cost, aesthetic appeal and energy efficiency.

Working hand in hand with the Department of Energy, UT-Battelle embarked upon a long-term effort to eliminate expensive marginal space that would be replaced with modern new facilities. Integrating such an ambitious goal of demolition and construction would require an unprecedented level of creative planning, political support, and operational discipline over an extended period. While a portion of the story warrants more detail, ten years later UT-Battelle had exceeded even its own expectations. More than 500,000 square feet of inefficient space had been removed, replaced with some 1.3 million square feet of modern and cost-efficient facilities that were attractive to prospective employees and an asset to growing the Laboratory's research program. Perhaps most significant, by 2012 an amazing 48 percent of ORNL's facilities were less than ten years old. The modern new facilities set the table for an accelerated period of growth that ultimately doubled the size of the Laboratory's research portfolio.

Changing the Culture

In the years to come, UT-Battelle's early efforts to improve safety and reduce costs at ORNL will likely not receive the attention afforded other, more glamorous achievements in the Laboratory's research programs. Indeed, UT-Battelle's vision to remold and upgrade ORNL's research programs in neutron sciences, computing and biology was certainly a more compelling case to the scientific stakeholder community and central to the permission extended UT-Battelle to undertake aggressive operational initiatives.

That said, many of the extraordinary accomplishments in the research mission in the years to come were made possible by a fundamental cultural change that began during UT-Battelle's first year. The magnitude of the change helped ORNL staff to understand the Laboratory's potential, and over time to become genuinely excited about what the Laboratory could be.

Following page: *An annual drawing for a new truck was part of UT-Battelle's efforts to promote safety. A dramatic reduction in accidents was part of a cultural change at ORNL.*

Winter 2000

Breaking the Mold

*"Progress sometimes requires taking a risk.
UT-Battelle took that risk, and the result has been
a tremendous period of progress at ORNL."*

—Senator Bob Corker

If any doubt remained about the need to address ORNL's aging infrastructure, it was removed on April 1, 2000, when UT-Battelle officially assumed management control of Oak Ridge National Laboratory. The Laboratory's emergency responders sent word shortly before noon that a large section of brick had fallen near the entrance of the vivarium, a 50-year-old facility more commonly known as the "Mouse House" that was home to the genomics research program. The idea that UT-Battelle came dangerously close to a fatality on the very first day of the contract sent a shudder through the new management team.

The Greatest Single Challenge

The falling bricks were viewed by some as an omen, and helped convince UT-Battelle that ORNL's dilapidated infrastructure represented the single greatest obstacle to the Laboratory's long-term vitality. From the perspective of the new management team, a variety of seemingly unrelated operational challenges at ORNL had their origin in an infrastructure that

was among the oldest and least efficient in the DOE laboratory system. Salaries of ORNL researchers were below average in part because limited resources were drained by utility costs in old buildings that were among the highest in the DOE laboratory system. Obsolete facilities could not meet the demands of modern research for space, ventilation, power and vibration-free foundations. In a ferocious international competition for talent, much of ORNL looked shabby beside state-of-the-art research campuses at universities or in the private sector. The difficulty in recruiting world-class talent, combined with facilities that were expensive and frequently unsuited for complex research activities, left ORNL increasingly unable to compete successfully for major research programs.

To UT-Battelle, the consequences of this trend were inescapable. The impact of an aging infrastructure would not be sudden, but rather would be manifested in a gradual atrophy of the Laboratory's capabilities and research base. Unless reversed, at some point

ORNL would inevitably lose the critical mass of talent and programs required to sustain its viability as a multi-purpose, world class research institution.

The most intimidating aspect of ORNL's infrastructure problem was its sheer scale. Faced with constrained budgets, capital investments in renovations and new construction at many DOE laboratories had been ignored or deferred for decades. At ORNL the volume of capital needs had grown to the point that the annual amount allocated for capital projects in the DOE system was not enough to keep the Laboratory from falling even further behind. Getting DOE approval and

Previous page: *Oak Ridge National Laboratory, circa 2000. UT-Battelle was convinced that rusted fences, massive asphalt parking lots and outdated buildings shaped ORNL's image, both in Oak Ridge and in Washington.*

Left page: *Senator Bob Corker. A private developer before entering public service, Corker was supportive of innovative ways to build public buildings faster and cheaper.*

The Department of Energy's historic transfer of property for private and state construction at ORNL opened the door to a $300 million modernization project. From left, ORNL Director Bill Madia, Congressman Zach Wamp, Congressman Jimmy Duncan, Energy Secretary Spencer Abraham, Senator Fred Thompson, DOE-Oak Ridge Manager Leah Deaver.

congressional funding for a single building could take years. Everyone agreed the notion that enough congressional funding would somehow become available to solve the scope of ORNL's infrastructure problem was completely unrealistic. Not surprisingly, in the spring of 2000 most ORNL staff seemed resigned to a Laboratory infrastructure that was bad and that would not likely get much better.

UT-Battelle determined that rebuilding the ORNL campus—and doing it quickly—would be the Laboratory's highest operational priority and

one absolutely essential to the goal of revitalizing and growing the Laboratory's research program. As early as April of 2000, senior officials at the University of Tennessee and Battelle were engaged in extended discussions around a strategy officials believed could make a substantial dent in the infrastructure problem in five years and within a decade deliver a near-complete modernization of the ORNL campus. The strategy would draw upon some four decades of relationships by Battelle with the Department of Energy, and an even longer partnership between the University of Tennessee and the state's

federal funds with both state allocations and private capital. The idea of building DOE facilities with private funds on federal land had no precedent. The suggestion that a state legislature fund capital construction for research facilities at a federal laboratory was unheard of in both Washington and Nashville.

A Delicate Bargain

First presented to the Department of Energy in May, UT-Battelle's modernization proposal was a straightforward bargain. The state of Tennessee would be asked to fulfill its commitment with $26 million to build on the ORNL campus three new facilities—to be owned by the state and shared with the University of Tennessee—that would support DOE's research missions in biology, materials, and high-performance computing. In return, the Department of Energy would commit to support the construction of six new facilities that included the Spallation Neutron Source, a modern laboratory for genomics research, and a multi-purpose building that would house conference space, a visitors center, and a modern cafeteria. The third and most creative piece of the proposal called for the construction at ORNL of three new facilities paid for by $72 million in private capital on land formerly owned by the Department of Energy. Taken collectively, the proposal's components represented ten new buildings with a total value of approximately $1.8 billion and an accelerated time table estimated at 7–9 years.

UT-Battelle's proposal was as delicate as it was novel. Each facet of the funding plan was contingent upon the success of the others. Viewing ORNL as untapped potential for the state's economic development and

the University of Tennessee's research program, Governor Don Sundquist reiterated his earlier pledge and recommended to the Legislature an initial sum of $8 million to fund a new facility to house the Joint Institute for Computational Sciences.

Supported by the University of Tennessee, the renewed commitment by state government to fund facilities at ORNL provided momentum for the second unprecedented initiative. To obtain private financing, UT-Battelle would have to convince DOE to transfer to private ownership a six-acre parcel located literally in the middle of the ORNL campus. Such a transfer would have to overcome, in addition to decades of institutionalized government policy, serious reservations about how facilities on private property would be constructed and managed, as well as how they could be expected to support DOE's mission on an ongoing basis. Absent a willingness by DOE to reverse decades of policy, ORNL's modernization plan had little chance of success.

The answer came in September 2000 when Energy Secretary Bill Richardson traveled to Oak Ridge and publicly endorsed UT-Battelle's modernization plan. Standing beside Governor Sundquist, Richardson unveiled a drawing of the envisioned ORNL campus, complete with federal, state, and private facilities. The Secretary's endorsement provided a critical signal to the financial markets that the project was real. Lending further support to what was no longer exclusively a federal project, state government made available resources to help complete environmental assessments and monitor construction of private and state-funded buildings. DOE attorneys joined their UT-Battelle counterparts in drafting stacks of legal documents

political leadership. More significantly, the plan would require that UT-Battelle and DOE sail into legal and financial waters that were largely unchartered. For everyone involved, the stakes and the risks were high.

UT-Battelle's modernization plan was based upon the simple premise that because congressional appropriations alone would never be adequate, funding from multiple sources would be necessary. Translated, the rapid construction of as many as a dozen new buildings would require an unconventional strategy to bolster

needed to transfer the property and finalize the lease agreement.

Significantly, the transition in Washington from a Democratic to a Republican administration after the November 2000 elections did not slow modernization plans. DOE's new leadership embraced UT-Battelle's efforts to use creative approaches to funding and constructing facilities in support of DOE's mission. In June 2001, Energy Secretary Spencer Abraham came to Oak Ridge to sign the deed transferring DOE property to the newly created UT-Battelle Development Corporation. Several hurdles remained, but for the first time even skeptics began to believe that UT-Battelle's plan might actually become a reality.

Overcoming the Unexpected

With quiet support from DOE's Oak Ridge office, implementation of the modernization plan accelerated throughout the summer. Working for the first time outside the government's conventional process for procurement and construction, ORNL management, through the UT-Battelle Development Corporation, modified and fast-tracked the process of selecting a firm to design and build the three privately funded facilities. Just as in the private sector, the flexibility gave UT-Battelle the opportunity to solicit comprehensive proposals for design, funding, scheduling, and construction based on the concept of "best value." The new approach attracted seven proposals, each with distinctly unique architectural and financing components. The project's new-found flexibility was reflected in the winning proposal, which urged UT-Battelle to reduce construction and operating costs by combining the three facilities into one, begin design work prior to the bond closing, and accelerate construction by starting

when 30 percent of the design was completed. In a conventional process, such changes might have taken months or even years to gain necessary approval.

Less than three weeks after the selection of the contractor, UT-Battelle's modernization plan was threatened by the kind of challenge that no one could have anticipated. The September 11 attack on the World Trade Center sent a shock wave through the nation's financial markets. The anticipated bond insurer for UT-Battelle's ORNL project suffered extensive human and financial losses in the attack and had to withdraw from the deal. Overnight, multi-million dollar bond ventures such as UT-Battelle's were viewed on Wall Street with a greatly heightened sense of caution. The unconventional nature of the ORNL project, which involved a private developer building and leasing a large research facility, occupancy by a third party, and funding contingent upon future appropriations from the federal government, served only to intensify the reservations of the bond market.

Unfortunately for UT-Battelle, the repercussions from the terrorist attacks were followed in October 2001 by another unforeseen series of events that further clouded efforts to finance the project. A succession of corporate scandals, many of which involved deceptive accounting and financing schemes, served to make bond analysts even more skeptical of proposals that included unusual elements. ORNL managers found themselves pitching one of the most unique and complex proposals in the history of government construction at the very moment investors were seeking simplicity and a minimum of risk.

Discussions with potential investors throughout the winter made evident

that the bonds would have to be sold at a higher cost if the project was to survive. Finally, on March 4, 2002, Bank of America closed a $72 million bond sale that provided the needed funding. Unexpected events had delayed the project about three months. But within two weeks of the bond sale, workers began tearing up parking lots and digging foundations for the largest construction effort in Oak Ridge since the Manhattan Project of World War II. Sailing into the wind, UT-Battelle was leading the Laboratory on a distinctly new pathway to research.

By the end of 2005, the Laboratory's physical transformation was becoming a reality. The massive parking lot had been removed, replaced for the first time by a large green lawn, or "quad," that formed the center of the new East Campus. Around the quad were three major new facilities—one funded by DOE, one by the state of Tennessee, and one by private capital—with a single architectural theme and that together represented the new spirit of UT-Battelle and the new future for ORNL.

At UT, at Battelle and at the Laboratory, the success in replacing much of the Laboratory's infrastructure in the face of enormous obstacles brought officials a newfound confidence. In their first years as new managers, no single achievement was of greater importance than breaking the code of modernization. Suddenly, ORNL's potential seemed unlimited.

Left page: *Construction of the Center for Computational Sciences, 2002. Financed and managed by the private sector, the $72 million facility was the innovative centerpiece of UT-Battelle's modernization strategy.*

Following Page: *Oak Ridge National Laboratory, circa 2007. In fewer than five years, a unique combination of federal, state and privately-funded facilities, complete with grass lawns and waterfalls, transformed ORNL into a modern research campus.*

Summer 2010

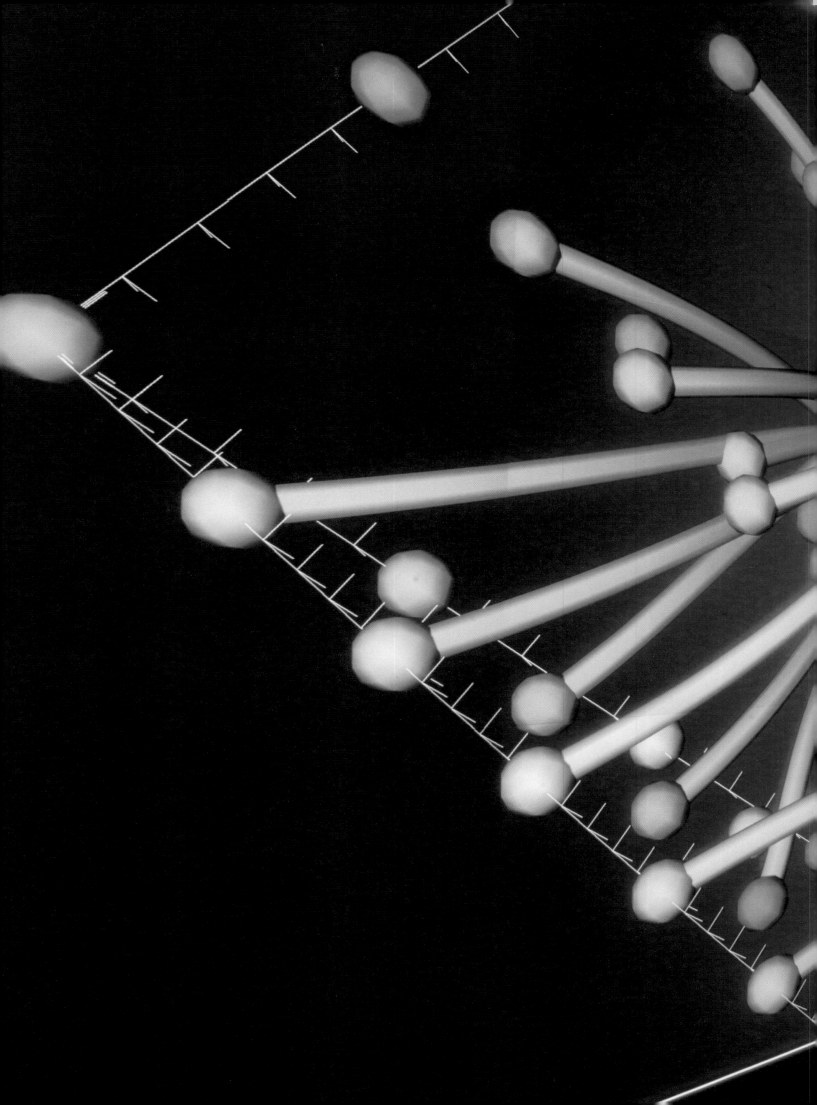

High-Performance Computing
ACHIEVING WORLD LEADERSHIP

"When the folks at ORNL told me they wanted to build a computational research program, they said they weren't satisfied with just being good enough. They wanted to be the best in the world."

—*Senator Lamar Alexander*

UT-Battelle's modernization plan for ORNL was not driven simply by a desire to beautify the campus or even to reduce costs. For the senior leadership at the University of Tennessee and Battelle, the campus modernization initiative was always part of a broader strategy to strengthen and expand the Laboratory's research agenda. The strategy was based on the belief that state-of-the-art facilities, in addition to attracting top-flight talent, could also be the key discriminator in efforts to prevail against leading universities and other DOE laboratories in the intense competition for new programs and increased research funding.

In no area was the modernization strategy more critical than high-performance computing. UT-Battelle, along with the Department of Energy, believed that future efforts to address "grand challenges" such as climate change or energy storage would require the integration of various scientific disciplines that combined experiment, theory and computational simulation. Equally significant, they believed that in each instance discovery would rest upon a foundation of computational power beyond the imagination of scientists only a generation earlier. From UT-Battelle's perspective, the laboratory that developed the world's foremost capabilities for high-performance computing would have a tool that represented an enormous advantage for tackling virtually every research challenge in DOE's portfolio.

Previous page: *Computer visualization. Using 35 megapixels on a 30 x 8-foot screen, ORNL scientists have the capability for real-time visualization and analysis of data generated by the world's most powerful supercomputers.*

Right page: *Senator Lamar Alexander. Former President of the University of Tennessee, Alexander was an early supporter of UT-Battelle's strategy of using high-performance computing as a foundation for research in other areas.*

Taking the Risk

UT-Battelle's decision in 2002 to devote extensive resources to developing a computational strategy was, at least in the minds of some, an extraordinary risk for a new contractor. In effect, UT-Battelle borrowed $72 million to build the nation's most advanced facility for computation with no program to put in it. Two acres of space in the new privately-funded building were dedicated solely to computing, with 30-inch pipes located under the floor for cooling and a new substation constructed to provide power. The new facility boasted 10-gigabyte-per-second connections to the ESnet and Internet2 networks, a scalable high-performance storage system for storing simulation data, and a disk subsystem that could transfer data at speeds greater than 200 gigabytes per second. UT-Battelle's gamble provided the infrastructure for a supercomputer a million times more powerful than machines operated only a decade earlier, based upon the belief that in the not-too-distant future the Department of Energy would make a major investment in the world of high-performance computing. Exactly when was anyone's guess.

The prediction proved correct, sooner than most expected. In the spring of 2003, the Japanese stunned the computational world with the announcement of the "Earth Simulator," an enormous machine capable of a staggering 40 trillion calculations per second. More powerful than the 20 largest American computers combined, the Japanese machine presented

A unique advantage. UT-Battelle's infrastructure investments in massive amounts of cooling and power were part of a strategy to make ORNL a choice location for a new generation of supercomputers.

KRAKEN

alarming implications for America's national security as well as efforts to sustain economic and scientific leadership. Confronted with this threat, the Department of Energy responded quickly. DOE announced a competition to build what the agency called a National Leadership Computing Facility, with some $500 million dedicated to building a new machine that would restore American leadership in high-performance computing.

Virtually all of America's major computational centers entered the competition. Located primarily in California, Pennsylvania, Illinois and Texas, many brought to the competition a reputation for computational science that surpassed ORNL. What they did not have was a facility—or more accurately, the infrastructure—required

to support a supercomputer of such unprecedented scale. UT-Battelle understood in 2002 what was then not fully recognized within much of the computational community—that in the future the greatest obstacle to more powerful machines would no longer be the architecture, but rather the vast infrastructure of power, cooling and connectivity required to support the scale of computing that DOE envisioned. While many of ORNL's competitors had the necessary talent and experience to build and operate the new machine, none had a place to put it. The design and construction of a new building, and in most cases the infrastructure to support it, would take years and siphon off a substantial portion of DOE's available funds.

Meanwhile, UT-Battelle for two years had been preparing for such an opportunity, quietly stockpiling a collection of the nation's finest experimental researchers and theorists while putting in place the connectivity critical to working with other laboratories and universities. Led

by Thomas Zacharia, a joint faculty member with UT and Associate Lab Director for Computational Sciences, ORNL submitted a proposal that included a modern facility equipped on day one to design and build the DOE machine. As part of UT-Battelle's competition strategy, the proposal was leveraged by the University of Tennessee's Joint Institute for Computational Sciences, under construction across the street at ORNL. On May 12, 2004, Energy Secretary Abraham surprised much of the computational community by announcing that Oak Ridge would be the site of the National Leadership Computing Facility. Boasting an output more than double that of the Japanese Earth Simulator, the enormous new machine, named "Jaguar," would be designed to produce a capability of 100 trillion calculations per second, with an anticipated upgrade in 2009 to more than 3000 trillion calculations.

DOE provided clear direction.. Models and simulations on the supercomputer would offer scientists a "third pillar of

science," a transformational addition to the historic pillars of theory and experiment. Equipped with a new generation of software and operated with unprecedented standards of efficiency, Jaguar would enable researchers to explore biology, chemistry, and physics in ways previously unimaginable. With the announcement, the world's center of gravity for high-performance computing began shifting for the first time to Tennessee.

For UT-Battelle, the pieces of the strategy developed in early 2000 were coming together. At its core, the strategy sought to combine modern new facilities with an unmatched computational capability to establish world leadership—and generate substantially increased funding—in a number of key research areas. Still under construction, the Spallation Neutron Source promised to provide a premier center for materials research. A second new facility had made possible the design of the world's most powerful electron microscope. In yet another

new building, the capabilities of the world's most powerful supercomputer laid the groundwork for research partnerships with major computational players such as the National Science Foundation, National Oceanographic and Atmospheric Administration, and the National Aeronautics and Space Administration. In time, ORNL's unique computational capabilities attracted a variety of leading industries interested in product simulations too expensive for conventional testing. As ORNL researchers increasingly worked with companies such as Boeing to develop lighter aircraft and partnered with universities such as Dartmouth to perfect new forms of biofuels, the Laboratory's brand and reputation increased in value.

Leveraged by the University

UT-Battelle's grand strategy also involved leveraging the new ORNL facilities for its UT partner. The excess capacity of the nation's most sophisticated computational infrastructure presented a chance for

the University of Tennessee to take advantage of an opportunity available only to a handful of the world's research universities. Like DOE, the National Science Foundation determined that developing high-performance computational capabilities would be a critical part of the Foundation's future research agenda. In 2007, NSF sought proposals to build a machine with a capability in excess of 800 trillion calculations per second. The solicitation produced responses from the nation's leading computation research centers in San Diego, Pittsburgh, Austin, Champaign, and Berkeley.

Historically, NSF had done little work with the federal laboratories, preferring to allocate the large majority of resources to academic research programs. Recognizing the NSF history of academic partnerships, the University of Tennessee developed a creative proposal that featured joint faculty at UT and ORNL who would conduct groundbreaking research on an NSF supercomputer supported by ORNL's unmatched infrastructure. Once again, ORNL's modern facilities proved to be the discriminator that tipped the balance. In the spring of 2008, the NSF announced an award of $65 million to build a massive machine operated by the University of Tennessee, the largest single research award in the University's history.

The new computer, named "Kraken," came on line in February 2009 with a capacity of more than 600 trillion calculations per second. By the end of the year, Kraken had exceeded 1,000 trillion calculations per second, giving the University of Tennessee the largest academic computer in the world. Seemingly overnight, UT had surged from the back of the pack to a place among America's elite computational research centers.

By November 2009 Oak Ridge National Laboratory and the University of Tennessee had, in a period of less than five years, developed two of the world's three largest machines. Jaguar, at 3,000 trillion calculations per second, or 3.0 petaflops, was number one. Kraken at 1.1 petaflops was number three. Meanwhile, the Laboratory was hiring some of the world's foremost computational scientists at the unheard of pace of one per week to conduct research on Jaguar and Kraken. While the expansion of machines and computational power around the world continued at a mind-boggling pace, Oak Ridge could lay legitimate claim to being the world's leading center of high-performance computing, having surpassed Japan, China and Germany as well as the traditional computational centers in America.

Delivering the Science

As much as ORNL and UT relished the visibility and acclaim of having the world's largest supercomputers, both

institutions recognized that the only real value of the massive machines rested in the volume of cutting-edge research they made possible. Kraken was funded by the National Science Foundation primarily for use by the academic community. Jaguar had a different mission that focused on a relatively small number of high-impact projects determined by DOE to be of national importance, with users chosen through a peer-reviewed competition from proposals submitted from industry, academia and government agencies. The machines were dedicated to unclassified research, much of which could not have been conducted previously on smaller machines.

Both Jaguar and Kraken undertook research projects—many energy related— involving tens of millions of processor hours and that previously were impractical due to the time required or the limitations of computer capacity. Jeremy Smith, a biophysicist who holds a Governor's Chair and a joint UT-ORNL appointment, was awarded 25 million processor hours to run simulations to help reveal the inner workings of lignocellulosic biomass, a raw material for biofuel production. Anthony Mezzacappa, another UT-ORNL joint faculty member, led a team awarded 34 million processor hours to develop three-dimensional models of core collapse supernovae. A third joint appointment, computational chemist Robert Harrison, was awarded 75 million processor hours to investigate the design of catalysts for the development of new manufacturing processes, including hydrogen production and storage. The fact that most of these projects were competitively awarded was evidence that UT-Battelle's investments in new talent and new facilities were producing tangible results for the research programs of both the Laboratory and the University.

Perhaps the most significant aspect of UT-Battelle's computational strategy was a commitment to sustain the Laboratory's leadership in one of the world's most competitive and rapidly changing technologies. By 2011, Jaguar's international ranking had dropped to third among the most powerful supercomputers on the Top 500 list behind Japan's K computer, with a record-setting performance of 8.162-petaflops, and China's Tianhe-1A, with a peak performance of 2.331 petaflops. The dominance of the Asian computers was short-lived. By late 2012, UT-Battelle announced the delivery of Titan, a new generation of supercomputer, twice as fast and three times as energy efficient as Japan's K computer. The Titan system was projected to have a peak performance in excess of 20 petaflops of high performance computing power, once again setting the standard by which the world's greatest computers are measured. As the Titan machine came on line, UT officials were already developing proposals to NSF to upgrade and expand Kraken.

Some twelve years had passed since the initial decision by the University of Tennessee and Battelle to take one of the most ambitious risks in the Laboratory's history. As a result of that risk, world leadership in high-performance computing had returned to the United States, and the research programs of ORNL and UT had set off on a path that few had thought possible.

Thomas Zacharia and Lee Riedinger. Both served as Deputy Director for Science & Technology at ORNL, Zacharia from the Lab and Riedinger from the University.

Neutron Science

"The research community is just beginning to fully appreciate the amazing capabilities of the Spallation Neutron Source. They are discovering that its potential to improve our lives is almost unlimited."

—Thom Mason, Director, Oak Ridge National Laboratory

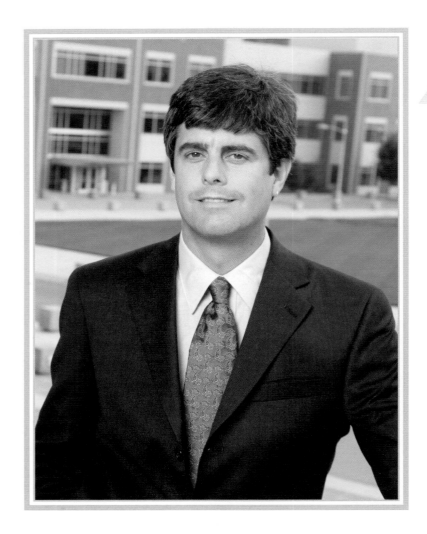

Dr. Thom Mason, Laboratory Director, 2007–present.

The successful effort to win the National Leadership Computing Facility in 2004 represented only a part of UT-Battelle's strategy to expand ORNL's research portfolio. In addition to securing a position of world leadership in high-performance computing, the strategy also included a commitment to reestablish ORNL's historical reputation as a premier center for neutron science and the study of materials. Indeed, the Department of Energy viewed this goal, with its direct relationship to the completion of the Spallation Neutron Source, as one of UT-Battelle's primary performance expectations during the Lab's initial five-year contract.

The science of neutron scattering was born in Oak Ridge at the Graphite Reactor, built in 1943 under the supervision of Enrico Fermi for the Manhattan Project at the site that became Oak Ridge National Laboratory. The reactor's original purpose was to demonstrate the production of fissionable plutonium, laying the groundwork for development of the atomic weapons that ended

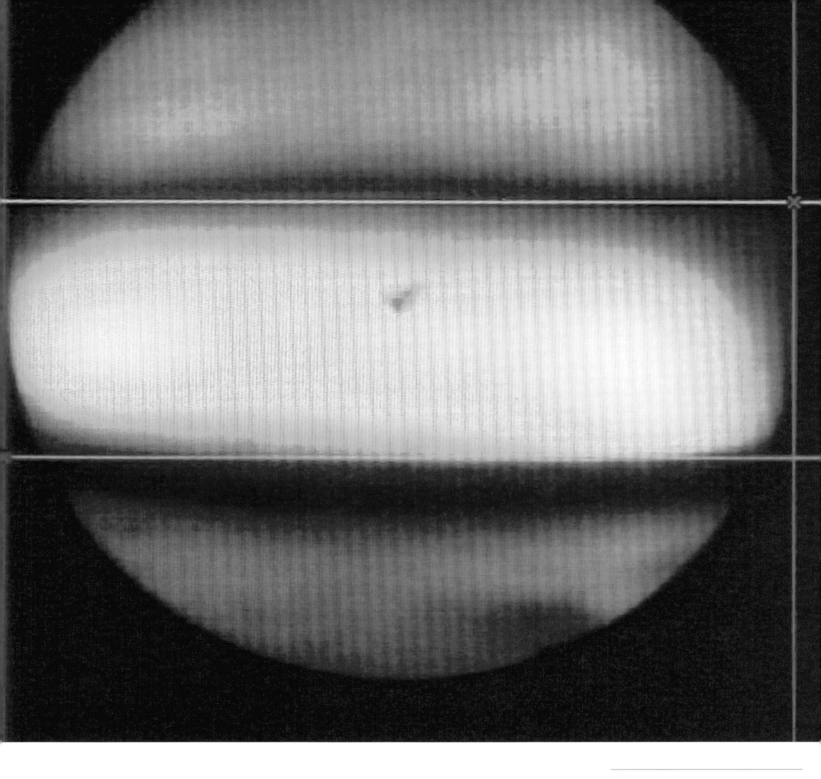

World War II. With its wartime mission completed, the reactor became a uniquely powerful tool for exploring the structure of materials. ORNL physicists Ernest Wollan and Clifford Shull led the development of neutron scattering techniques and their application to studies of solids and liquids—work ultimately recognized with a Nobel Prize awarded to Shull in 1994 after Wollan's death. For almost three decades, the United States led the world in neutron science, a leadership based in part on research reactors at Oak Ridge National Laboratory. In the 1970s, the U.S. surrendered the leadership role to Europe, where a new and larger generation of neutron sources was built in France and Great Britain.

Delivering the Promise

Since UT-Battelle's success in securing an exemption from the state sales tax for SNS construction in January 2000, moving ahead with the construction of the world's largest science project had become DOE's most important

Above: *Neutron produced by the Spallation Neutron Source. The ability to understand the alignment of neutrons enables scientists to develop new materials that are stronger, lighter and more durable.*

Previous page: *Storage pool, High Flux Isotope Reactor. At 85 megawatts, the upgraded HFIR is one of the world's largest research reactors, supporting nuclear energy programs at both UT and ORNL.*

ORNL priority, both in Washington and in Oak Ridge. For DOE, delivering the $1.4 billion linear accelerator on time and on budget was a critical aspect of regaining American leadership in neutron science, providing a spectacular tool for the nation's economic competitiveness, and developing congressional confidence in the agency's ability to manage projects of such enormous scale and complexity. For UT-Battelle, the SNS project represented a chance to establish credibility with their DOE customer and—equally important—put in place an "anchor" scientific facility that would greatly enhance the Laboratory's research program, reassert ORNL's reputation in neutron science, and put to rest any notion that budget restraints might make ORNL a target for closure.

Virtually every aspect of the SNS project defied conventional wisdom about the process of designing such a massive scientific facility or securing comparably large congressional appropriations. To secure the breadth of expertise needed for the design of an extraordinarily complex machine with more than 100,000 separate and interdependent control points, DOE had adopted a new project management model that leveraged talent from across its laboratories, with the simultaneous goal of accelerating the project's technical execution while controlling cost. UT-Battelle inherited an SNS project that was being designed and built by six DOE laboratories—Argonne, Brookhaven, Jefferson, Lawrence Berkeley, Los Alamos, and Oak Ridge—operating under different corporate cultures and located thousands of miles apart. The laboratory collaboration was supplemented by a remarkable collection of talent and experience from 23 states and 15 countries, including Canada, France, Germany, Great Britain, Japan, Korea and Russia. The talent was needed. The SNS demanded beam lines so precisely aligned that the Earth's curvature was factored into the construction of the linear accelerator, a tiny but critical difference of 7 millimeters from one end of the 1000-foot linac to the other. All SNS accelerator and target components—independent of size, shape, and weight—were installed to specifications within 0.2 millimeter. Together, they would comprise a machine that used enough power to supply the 30,000 citizens of Oak Ridge.

Managing the project was not without challenges. When UT-Battelle was selected for the management contract at ORNL, some in the scientific community—perhaps envious of the large appropriations that would be required for the SNS construction—voiced concerns about changing the proverbial horse in mid-stream. Questions persisted in 2001 when UT-Battelle selected Thom Mason, a 36-year-old Canadian, to take over the SNS project in the same year Congress was being asked by DOE to provide an imposing $281 million in construction funds. He managed a work force that peaked at about 1,300 in November 2002, and included more than 3,800 employees over the seven-year span of the project.

Despite obstacles that many thought insurmountable, on April 28, 2006, a single computer key stroke fired ten trillion protons into the mercury

target of the SNS, releasing neutrons in a process that would help define American science for decades to come. With a pulse that eventually would be ten times more powerful than any comparable machine in Europe or Asia, the result was both tangible and symbolic. At that precise moment, the men and women who gathered around the SNS control room could look back with pride, knowing they were the heirs to the legacy of Clifford Shull and Enrico Fermi, and that they had restored to Oak Ridge—the birthplace of neutron scattering—the world's leadership in neutron sciences.

The Spallation Neutron Source was not ORNL's only facility that contributed to Oak Ridge's reputation as the world's undisputed leader for neutron science. The High Flux Isotope Reactor, at 85 megawatts, is the highest flux reactor-based source of neutrons for research in the United States. The steady-state neutron fluxes generated by HFIR are among the highest at any research reactor in the world and provide an ideal complement to the pulsed neutrons produced by the SNS. More than 500 researchers annually combine an unmatched collection of instruments with the thermal and cold neutrons produced by HFIR to study the fundamental properties of condensed matter, including the effects of radiation on materials. Research is not confined to the reactor. Spent fuel assemblies in the reactor pool provide a gamma irradiation facility used to study the effects of long-term radiation exposure on a variety of materials. HFIR is also one of America's leading centers for the production of isotopes for medical, scientific and industrial research.

Becoming a Player

UT-Battelle's aggressive agenda for the expansion of ORNL's capabilities in materials science was not limited to the Spallation Neutron Source and High Flux Isotope Reactor. In 2001, lab officials began plans to compete for one of five new nanoscale science research centers to be funded by the Department of Energy. The nanocenters were envisioned to provide the scientific community with state-of-the-art equipment and the research expertise necessary to support the emerging multidisciplinary areas of nanoscience and nanotechnology.

For UT-Battelle and ORNL, the significance of the competition for the nanocenters extended well beyond the opportunity to acquire new research capabilities. The competition was the first time since UT-Battelle assumed management control that ORNL had gone head to head with other DOE multi-program laboratories on a project of such scale and visibility. The competition would be a test not only of how well ORNL could prepare under pressure a proposal of the highest complexity, but also of how DOE viewed UT-Battelle's ability to handle another large construction project simultaneously with the Spallation Neutron Source.

ORNL's nanoscience proposal was coordinated in part by Lee Riedinger, the Laboratory's Deputy Director of Science and Technology, and Jim Roberto, Associate Laboratory Director for Physical Sciences. Riedinger came to ORNL from the University of Tennessee and previous roles as Head of the Physics Department and Director of the UT-ORNL Science Alliance. Roberto was a holdover from the previous contractor. Together, they produced a proposal that recommended a site immediately adjacent to the SNS and rested on ORNL's traditional strengths in materials synthesis and characterization as well as in theory, modeling and simulation, and included both "hard" and "soft" materials. Significantly, the proposal stressed ORNL's success in managing the cost and schedule milestones of the SNS project. The proposal also leveraged the state of Tennessee's commitment to build a new Joint Institute for Neutron Sciences, a facility for University of Tennessee students and faculty within walking distance of the proposed new Nanocenter.

In 2003, DOE selected ORNL to house the first of the agency's five nanoscale science research centers. Funded at $65 million, the 80,000 sq. ft. Center for Nanophase Materials Sciences contained a 10,000 sq. fit. nanofabrication clean room where researchers could create materials 10,000 times smaller than a human hair. The ability to develop new materials of unimaginable size and strength had unlimited potential for virtually every sector of the American economy. ORNL's selection to host DOE's first nanocenter had two consequences. The addition of the new nanocenter to the outstanding research capabilities of the Spallation Neutron Source and the High Flux Isotope Reactor further strengthened ORNL's reputation as a leader in materials science. For UT-Battelle, the victory over some of the nation's premier laboratories signaled that ORNL was once again a major player in the American scientific community.

They're big, but what can they do?

Not long after the Spallation Neutron Source began operation in 2006, a

Knoxville journalist summed up a common attitude among the public. "We know it's big, and we know we're for it, but what does it do?" The comment expressed succinctly one of the biggest challenges facing UT-Battelle. It is not enough to claim the world's best facilities for materials research. The real measure of performance is whether the magnificent machines can produce the quality of science worthy of their size and cost.

The SNS, combined with its suite of instruments, provides scientists with an unprecedented understanding of materials properties. The size and nature of the instruments are related to the specific material and properties that researchers wish to examine. One example is the lubricants used in a multitude of industrial products. By using neutrons to study lubricants that spread when a force is applied, researchers have discovered which additives improve fluid properties. Modern oils stick to moving metallic engine parts, whereas older oils, when heated during start-up, spread and separate from the parts they are intended to lubricate. With newly-designed lubricants, cars can generate more power with less emission. Bridges and other steel structures need to be painted less frequently. A new generation of tiny machines can be lubricated with thin-film coatings instead of oil-based fluids.

In a different category, each new generation of commercial and military aircraft is expected to travel faster and farther while using less fuel. To meet these demands, the aircraft must be made of lighter materials held together with stronger, lightweight welds rather than heavy rivets. Neutron-scattering results, combined with computer

models made possible by ORNL's enormous computational capacity, provide an enormous advantage to American industry by helping engineers develop materials and improve welding processes.

The spectrum of discoveries for new products is unlimited. Neutron research on soft matter could lead to time-released drug delivery systems that target specific body organs and release a medicine precisely when needed. Superconducting cables and components will reduce electricity costs and help revolutionize the American grid system by carrying far more power in a fraction of the space. Research at the Nanocenter is paving the way for advanced batteries for energy storage, high-efficiency solar photovoltaics, and lightweight fuel cells capable of powering emission-free vehicles. The application of polarized neutrons at the SNS will guide the development of tomorrow's quantum supercomputers. Diseases could be cured through a better understanding of how proteins work in the human body.

Seven decades have passed since Nobel Laureate Clifford Shull and his colleagues performed their ground-breaking neutron scattering experiments at the Graphite Reactor. For three decades, America led the world in neutron science before surrendering the lead to Europe. In Oak Ridge—the birthplace of neutron scattering—it is fitting that UT-Battelle and ORNL have restored that leadership to America.

Right: *Center for Nanophase Materials Science. Located adjacent to the SNS, the nanocenter is a major asset in ORNL's inventory of materials science capabilities.*

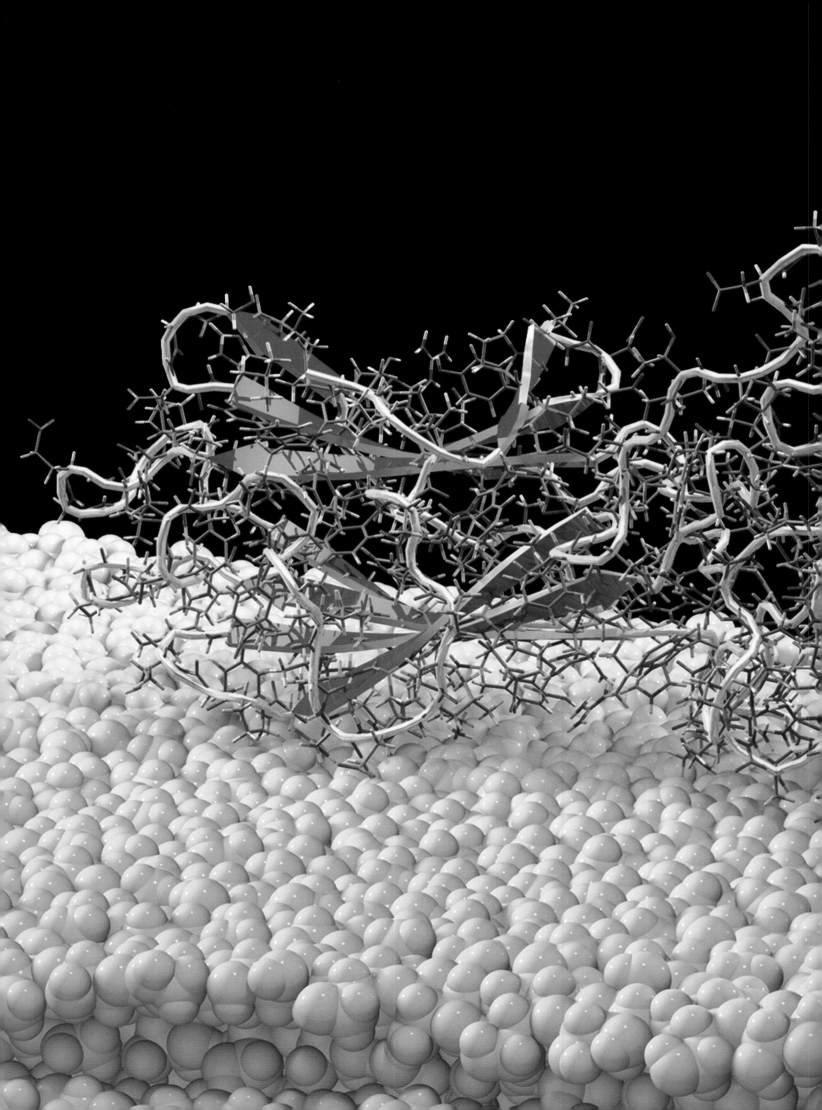

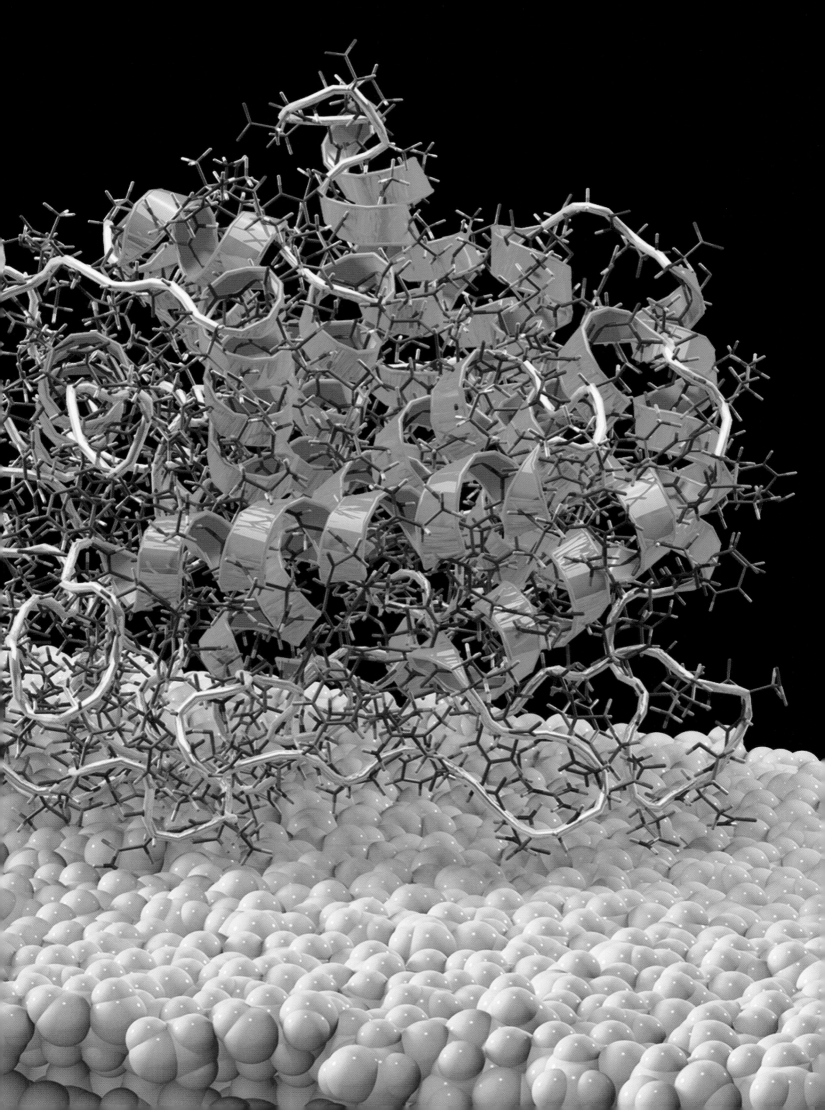

Energy

> ❝*Energy research is a great example of where our partnership made a critical difference for the University, for the Laboratory, and for the Department of Energy's mission.*❞
>
> —*Dr. David Millhorn, Executive Vice President, University of Tennessee*

Throughout the ups and downs over the years, one constant remained. Since the earliest days of the Manhattan Project, Oak Ridge National Laboratory was one of the world's premier institutions for energy research. Beginning with the harnessing of nuclear power at the Graphite Reactor in 1943, ORNL had developed a portfolio of energy research that was the largest and most diverse in the United States. ORNL's research programs were widely considered among the nation's leaders in nuclear energy, fusion, energy efficiency, transportation, bioenergy, electronics, sensors and energy distribution.

Arriving at ORNL in April 2000 with the fresh eyes of an outsider, UT-Battelle viewed the Laboratory's energy portfolio with a different perspective. While most of ORNL's energy programs could be considered as "among the best," none could claim undisputed world leadership. Without a concerted strategy to strengthen capabilities and target growth in one or more energy

sectors, lab officials worried that the long-term trend could be a gradual contraction of the energy programs and the dilution of ORNL's brand as a leader of energy research.

Among the first decisions was a commitment to reinvigorate ORNL's program of nuclear research. The decision was based upon ORNL's historic leadership in nuclear energy, the presence on campus of one of the world's largest research reactors, and the potential of faculty and student support from the University of Tennessee's Department of Nuclear Engineering, one of the nation's largest pipelines of talent for the nuclear industry. Like many facilities at ORNL, the High Flux Isotope Reactor was four decades old. By the summer of 2000 the research reactor was frequently forced to shut down for extended repairs at a substantial cost of time and money both to the Laboratory and to visiting scientists. With the nearby Spallation Neutron Source under construction, the scientific community

Previous page: Extraordinary computational power makes it possible for scientists to understand the process of breaking down cell walls to make biofuels.

began to wonder aloud if the Department of Energy might choose to close the aging reactor.

Instead of being resigned to HFIR's closure, UT-Battelle advocated upgrades to the reactor as part of the Laboratory's modernization plans. The Laboratory contended that HFIR, instead of being redundant to the pulsed neutrons of the SNS, in fact offered a unique complement source of steady-state neutrons that could provide a one-of-a-kind combination available at

only one site in DOE's nationwide complex of laboratories. The argument proved compelling. In 2007, HFIR completed a $62 million upgrade, the most dramatic transformation in the reactor's 40-year history. During a shutdown of more than a year, the refurbished facility installed a number of modern instruments, including a cold neutron source, new computer system controls, and eight thermal-neutron spectrometers in the beam room. The reactor was restarted in mid-May, attained full power of 85 MW within

two days, and resumed experiments within a week.

Perhaps as important as the upgrades was a new rigor of management brought to HFIR's operation. The number of unplanned outages, as well as the number of recorded injuries to

Above: *UT Executive Vice President Dr. David Millhorn. A biologist by training, Millhorn's appreciation for research played a major role in successful collaborations between UT and ORNL.*

staff, dropped dramatically. The result was a restored confidence among the scientific community in both HFIR's capabilities and operational predictability. Responding to DOE's outreach goal, the number of users doubled, with more than 500 scientists from around the world coming each year to HFIR to conduct experiments.

Shortly before the completion of the HFIR upgrade, ORNL was called upon by the Department of Energy to take a leadership role in another energy sector, this time managing the U.S. contribution to ITER, a global initiative to build the world's most advanced magnetic fusion experimental facility in Cadarache, France. The ORNL team led the U.S. delegation in a multi-national $20 billion project that includes the European Union, China, India, Japan, Korea, and Russia. The project's ambitious goal seeks to harness the power of fusion, with the dream of generating a limitless source of energy.

ITER had little precedent in international science. The project's technological challenges were equaled only by attempts to develop common strategies and expectations among such a diverse collection of scientists, cultures and political systems. Despite these challenges, the ITER project is considered by many as a major step that could provide a sustainable future energy alternative. For ORNL, the rewards are less tangible. While American firms supply many of the reactor's parts, relatively few ITER staff are based in Oak Ridge and the results of their efforts cannot be measured for at least a decade. In a larger context, however, ITER represents a strong vote of confidence by the American scientific community in ORNL's capabilities,

---

Left: *UT's biorefinery tests the commercial viability of biofuels developed at ORNL.*

and a willingness by the Laboratory to contribute once again to a greater national good in the pursuit of science.

Ultimately, the quality and reputation of any laboratory are measured by the ability to compete successfully for the largest scientific research projects. Some of the competitions sponsored by the Department of Energy attract proposals from the best talent and the most prestigious government, academic and private sector research institutions in America. UT-Battelle's "breakout win" came in 2005 with DOE's selection of ORNL to host the new Leadership Computing Facility. If UT-Battelle was serious about reasserting ORNL's leadership in energy, it would require winning two of the most competitive and high-stakes competitions of the decade.

The Department of Energy's research mission is derived from the unique capabilities of their laboratories. The agency identifies what are often termed "grand scientific challenges" of importance to the nation's health or economic prosperity. In many instances, these scientific challenges require a large investment in fundamental research that makes the project unattractive to the private sector. No corporation, for example, could afford to build a Spallation Neutron Source or a petaflops computer to conduct research on a product years away from the market. The role of building these enormous facilities and funding large research projects fills a critical gap for American industry and has evolved into one of DOE's primary functions. DOE facilitates the process by soliciting proposals from laboratories and universities, which more often than not partner in teams built around specific capabilities. The largest proposals exceed $100 million, and attract teams that frequently represent the

world's best talent in a particular field of research.

In 2006, the largest and most visible competition was DOE's solicitation for three Bioenergy Science Centers. Funded at $125 million each over five years, the centers reflected DOE's long-term commitment to developing a more sophisticated and cost efficient generation of biofuels capable of replacing a substantial portion of fossil fuels used in the American auto market. UT-Battelle viewed the competition with mixed feelings. Winning such a high profile competition would greatly strengthen the Laboratory's reputation in an emerging energy sector. Conversely, losing could reinforce doubts about ORNL's ability to play among the nation's premier research institutions. One compromise option debated internally was whether to join another team as a supporting member.

UT-Battelle's decision to lead a proposal team for the Bioenergy Science Centers was a calculated risk consistent with the aggressive philosophy that characterized earlier initiatives. ORNL's 17- member team, in addition to four universities and three companies from the private sector, included major roles for the University of Tennessee and the state of Tennessee. Since competitions are evaluated in part on the extent to which proposals offer matching resources, the unique commitments from UT and the state provided what many hoped would set ORNL apart from dozens of other first-rate proposals. Indeed, the ORNL proposal offered to leverage DOE's investment with commitments of space and funding that far exceeded conventional "in-kind" resources involving faculty or graduate students.

Similar to the successful computational proposal strategy a year earlier, UT and

ORNL offered to make available the new state-funded Joint Institute for Biological Sciences, scheduled to open concurrent with DOE's Bioenergy Science Center. A brand new $12 million facility ready to go on day one was appealing to those who wanted to save time and money and stand the program up immediately. An already compelling proposal was strengthened further by UT's ability to make available another new facility dedicated solely to biofuels research. Located about 45 minutes from ORNL, a new $70 million state-funded facility capable of manufacturing 250,000 gallons of biofuels a year would be dedicated to testing the commercial viability of biofuels developed in the new research center.

Once again, UT-Battelle's proposal offered a rare collaboration of state and federal resources to deliver DOE's research mission. On June 26, 2007, DOE announced that the new Bioenergy Science Center would be located on the ORNL campus in a facility owned by the University of Tennessee and funded by the state of Tennessee. Employing the interdisciplinary expertise of the team's partners in biology, computation, engineering and agricultural science and commercialization, the ORNL team met and exceeded most of DOE's milestones to develop revolutionary processes for converting plants into fuels. For ORNL, a victory over 20 of the nation's strongest teams, followed by the delivery of a challenging set of expectations, was an enormous boost to the Laboratory's reputation as a leader in energy research.

UT-Battelle barely had time to celebrate before turning attention to another energy-related competition, this time focused on the American nuclear industry. Nuclear power supplied twenty percent of the nation's electricity, made possible by an aging

fleet of reactors that was rapidly approaching their life expectancy. The Three Mile Island accident in 1979 resulted in a virtual moratorium on the construction of new reactors. Confronted with an aging nuclear fleet and the inability to build new ones quickly, the Department of Energy was seeking creative technologies that would enable existing reactors to operate longer and provide new materials for the next generation of reactors.

Named the Consortium for Advanced Simulation of Light Water Reactors, or CASL, DOE's proposed new "Energy Hub" program was aligned with ORNL's capabilities and with UT-Battelle's strategy to reinvigorate the Laboratory's nuclear research program. UT-Battelle approached the competition with a simple template that was proving increasingly successful for both large and small projects. At its core, the strategy involved providing state-of-the-art facilities, gathering the best available talent from ORNL, other laboratories and universities, and developing a multi-disciplinary proposal that took advantage of ORNL's world-class capabilities in computation and materials.

UT-Battelle proposed to provide computational models generated in the world's most powerful computer, combined with data from materials tested in the world's largest linear accelerator. The program would be housed in 9,000 sq. ft. of new space, and supported by a team that included experts from four national laboratories as well as MIT, Georgia Tech, North Carolina State and Michigan. On May 28, 2010, DOE once again selected an ORNL team as host for the $122 million project that would extend the life of existing power plants, reduce the volume of spent nuclear fuel generated

in current reactors, and develop new designs for the next generation of light water reactors.

While ORNL's success in two major competitions in three years had revitalized the Laboratory's energy agenda, the growth was not confined to the nuclear and bioenergy programs. The transportation portfolio had grown to more than $100 million. New partnerships, including Nissan and the state of Tennessee, spurred activity in battery storage and electric vehicles. Another partnership with the Tennessee Valley Authority designed low cost homes that literally produced more energy than they used. The need to expand the capacity and security of the nation's electric grid emerged as a new research focus.

Individually, the Laboratory's energy programs represented a healthy research agenda that was responsive to the needs of the DOE customer. Taken together, the upgrade of the High Flux Isotope Reactor, combined with winning the Bioenergy Science Center and the CASL nuclear hub, were evidence that Oak Ridge National Laboratory had reestablished its reputation as the world's premier energy research institution.

Right page: *Computer simulation of fuel burned in a nuclear reactor. The research is designed to extend the life of America's aging nuclear fleet.*

Joint Institutes & Governor's Chairs
INDISPENSABLE ASSETS

> ❝ *We realized that the University and the Laboratory are critical components linked together in Tennessee's economic strategy.*❞
> —University of Tennessee Knoxville Chancellor Jimmy G. Cheek

Right page: *UT Knoxville Chancellor Jimmy G. Cheek. His support of the Governor's Chairs program is a major contribution to UT's goal of being a Top 25 public university.*

Previous page: *The state-funded Joint Institute for Biological Sciences, a critical factor in the Department of Energy's selection of ORNL to host the Bioenergy Science Center.*

At its heart, UT-Battelle's goal of strengthening ORNL's core capabilities and growing the Laboratory's research portfolio was tied to a simple strategy—provide state-of-the-art facilities and stock the facilities with some of the best talent available. The strategy was not original. While most national laboratories and research universities have some variation of the same theme, the goal is more easily proclaimed than accomplished. For ORNL, the most distinctive feature of this strategy, and the part that was absolutely critical to much of the progress enjoyed during UT-Battelle's first decade, were the unprecedented roles played by the state of Tennessee and the state's land-grant institution, the University of Tennessee. The extent of the partnership among a federal facility, a state flagship university and three administrations of state government were unique in the American research community. The fact that Tennessee's contribution to this partnership exceeded more than $150 million was extraordinary.

Most of DOE's national laboratories have university collaborations, including six laboratories that are managed or co-managed by universities. Indeed, Lawrence Berkeley National Laboratory is often viewed as a model of integration with some 500 joint faculty sharing time between the Laboratory and the University of California at Berkeley. At none of these laboratories, however, has the level of state investment approached the breadth and scale as Tennessee's investments in the UT-ORNL partnership. Why a relatively small state decided to make such a major investment in a federal facility—and

the returns that have come as a result of those investments—are central to understanding the resurgence of Oak Ridge National Laboratory and the University of Tennessee's role as an emerging research institution of national prominence.

In Tennessee, the University of Tennessee and Oak Ridge National Laboratory are among the state's most highly-valued economic assets. Comparable laboratories and universities in much larger states such as California, New York and Illinois occupy a less prominent role in the mix with some of the world's largest cities and corporations. With relatively limited assets, Tennessee must maximize each to succeed against bigger states in the competition for new corporate investment and economic expansion. Although UT and ORNL had worked together off and on since World War II, the relationship in 1999 had never approached the potential that some envisioned for two institutions only twenty miles apart. The leadership at both Battelle and the University of Tennessee realized that achieving this potential would require more than cosmetic changes in current programs. What they had in mind was a fundamental breakthrough that would view the University and the Laboratory not as distinct state and federal assets, but as critical components linked together in Tennessee's future economic strategy.

To facilitate the strategy, leaders at UT-Battelle, the University and the state reached a simple agreement. The agreement, never formalized, said that state government, working through the University of Tennessee, would make investments in facilities and personnel intended to leverage ORNL's ability to compete for new programs and expand the Lab's research base. The Laboratory in return would share its equipment and facilities with an expanded program of joint faculty and joint research at the University of Tennessee. In addition, UT-Battelle committed to taking down the fences, both figuratively and literally, in a concerted effort to turn technologies developed in the Laboratory into new companies and new jobs in Tennessee.

The strategy was elegant in its simplicity. ORNL would get three new buildings as part of the Laboratory's accelerated modernization plan. University of Tennessee faculty and graduate students would get expanded access to some of the world's most advanced research facilities. With a "two-for-one investment," the state of Tennessee would strengthen two of its most important economic assets and realize the potential for increasing numbers of high-tech start-up companies.

Three Joint Institutes

With the support of Governor Don Sundquist and the House and Senate Speakers, the state committed $26 million (a figure that was later increased) to construct three facilities on the ORNL campus that would be owned by the University of Tennessee and operated as joint institutes with the Laboratory. The joint institutes would house faculty and students from the University and serve to leverage ORNL's attempts to compete for larger federal research programs in computational, biological and neutron sciences.

Joint Institute for Computational Sciences

The first of three new facilities, the Joint Institute for Computational Sciences is located across the street from ORNL's Center for Computational Sciences on land deeded from the Department of Energy to the state of Tennessee. The 52,000 sq. ft. facility was a central feature in ORNL's 2005 proposal to the Department of Energy for the Leadership Computing Facility and in UT's proposal to the National Science Foundation in 2009 for the world's largest academic supercomputer. In 2010, the two Tennessee machines—Jaguar and Kraken—were the world's first and third most powerful supercomputers, both capable of a "petaflops," or more than 1,000 trillion calculations per second. Having established computational "critical mass," they were joined by Gaea, a third petaflops machine funded by the National Oceanic and Atmospheric Administration and dedicated to providing climate modeling simulations. In a period of less than five years, Oak Ridge, including the UT-ORNL partnership, had become the international center of high performance computing.

The opportunities for both faculty and students to work on machines of such extraordinary power were unmatched. At both ORNL and UT, joint faculty appointees, postdoctoral fellows, graduate students, and administrative staff participated in a conscious effort to use high performance computing as the foundation of multi-disciplinary research. The Joint Institute has helped develop interdisciplinary research programs in computational biology that use supercomputers to sift through trillions of letters of DNA and protein sequences, computational mathematics that assists the aerospace and automotive industries with cost-competitive designs, and chemical processes ranging from the molecular to the nanoscale.

Joint Institute for Biological Sciences

Established in 2007, the Joint Institute for Biological Sciences, perhaps more than any other state investment, exemplified UT-Battelle's strategy of leveraging state facilities as part of major research proposals. The investment included a new $12 million facility located on ORNL's west campus, once again on a parcel of federal property deeded to the state of Tennessee. Reflecting the risk at times involved in UT-Battelle's modernization and proposal strategy,

the facility was programmed and designed prior to DOE's announcement of the competition for the $125 million Bioenergy Science Centers that would help develop a new generation of biofuels. Both UT and ORNL immediately believed the opportunity to become a national center for bioenergy research justified a commitment to reprogram the new facility.

The decision had enormous dividends. With additional leverage from the University of Tennessee's new biorefinery, ORNL, along with Lawrence

Berkeley National Laboratory and the University of Wisconsin, were selected by DOE to host the three Bioenergy Science Centers. DOE's challenge for the ORNL team was straightforward—develop a cost-effective way of overcoming ten million years of nature's resistance to the breakdown of cell walls. Responding to the challenge required precisely the kind of multi-disciplinary approach favored by UT-Battelle, one that employed large scale simulations on the world's foremost neutron scattering facilities, systems biology data to help

understand microbes, and ultra-scale computers that can visualize on a 30-foot screen how simulated enzymes can break down cellulose strands.

The results have exceeded expectations, in some instances ahead of DOE milestones. Seeking to use switchgrass as an alternative to corn as a source for biofuels, scientists have discovered two exceptional switchgrass lignin genes with the ability to reduce the volume of lignin produced by switchgrass, increase its ethanol production and lower the amount of enzyme needed to sustain the process. They are striving to lower the cost of biofuels production by demonstrating that a modified strain of yeast can reduce biomass to sugars and ferment

---

the sugars into biofuels in one step. Most important, they have delivered one of DOE's primary milestones by using molecular biology to make plant cell walls easier to break down into their component sugars. Researchers say without exaggeration that a number of the most significant steps on the road to expanding America's non-fossil fuel options have taken place in Oak Ridge.

Joint Institute for Neutron Sciences

Unlike its two counterparts, the Joint Institute for Neutron Sciences had a mission that shifted with events. The facility was first proposed to the Governor in the early 1990s as a means of leveraging federal support, first for the Advanced Neutron Source, and later for the Spallation Neutron Source. When Congress appropriated funds to begin construction of the SNS in 2000, UT-Battelle's new management team elected to defer building the joint

institute because of concerns about potential confusion from a state and federal project under construction simultaneously. Depending upon one's perspective, the delay presented an unexpected opportunity to leverage the state facility for another major DOE proposal, the $62 million Center for Nanophase Materials Science, awarded to ORNL in 2003.

Owned by the University of Tennessee and constructed during the administration of Governor (and physicist) Phil Bredesen, the Joint Institute for Neutron Sciences serves as a gateway for users of the Spallation Neutron Source, the High Flux Isotope Reactor and the Nanocenter. Located within walking distance of the SNS, the Institute offers support to hundreds of domestic and international users who come to Oak Ridge each year to conduct experiments. The unique scientific data provided by the experiments include atomic and spin structure and dynamics in condensed matter physics, materials science, materials chemistry, polymer science, biological sciences, and engineering. The Institute has a particular focus on bringing graduate students and post docs from the University of Tennessee into the ORNL facilities, where they often conduct experiments under the instruction of some of the world's leading experts in neutron scattering.

Finding the Very Best

The second phase of UT-Battelle's strategy for the Laboratory's research agenda involved convincing some of the world's most outstanding scientists to come to East Tennessee. The primary targets were men and women with a uniform profile. They were at the top of their respective fields, attracted by the possibility of working on big challenges with state-of-the-art equipment in modern facilities, and invigorated by the opportunity to spend part of their time teaching and supervising university students.

Again, UT-Battelle enlisted the help of the state. As Chair of the UT Board of Trustees, Governor Bredesen appreciated the dual value to the University and to ORNL of developing a critical mass of exceptional talent that could strengthen the research agendas and reputations of both institutions. In 2005 the Governor pledged to make available $10 million to initiate what became known as the "Governor's Chairs," a program designed to hire top-tier researchers who would hold joint appointments with UT and ORNL. Within five years, the Governor's Chairs became one of the most valuable research assets at both the University and the Laboratory.

The first Governor's Chair was selected in 2006 and brought multi-dimensional capabilities in the field of computational molecular biophysics. He was followed by new appointments

in the disciplines of polymer chemistry, nuclear security, power electronics, energy storage, microbiology and environmental engineering, computational nuclear engineering, and computational genomics.

The Governor's Chairs produced the anticipated benefits. At the University, external research funding grew substantially, strengthening UT's goal of becoming a Top 25 public university. The chance to study under such outstanding faculty attracted an increasingly higher quality of undergraduates and graduate students at the University, and a comparable quality of post docs at the Laboratory. With critical help from state government, by 2010 UT-Battelle's strategy of building state-of-the-art facilities and stocking them with the best talent available had become a reality.

Below: *Seven Governor's Chairs recipients. Founded by Governor Phil Bredesen, the Chairs offer top scientists the chance to share time between UT and ORNL.*

> **"**In the laboratory and in the classroom, one of Tennessee's most exciting opportunities lies in combining the resources of a great university and a great national lab.**"**
>
> —*Governor Phil Bredesen*

A large portion of UT-Battelle's first decade at Oak Ridge National Laboratory was focused on laying the foundation for strengthening the relationships between the Laboratory and the University of Tennessee. Three new state facilities were constructed to house the joint institutes of computational sciences, biological sciences, and neutron sciences. The Governor's Chairs program, envisioned to provide joint appointments to the highest echelon of scientific researchers, was created and funded. The University and the Laboratory teamed to leverage these new assets and compete successfully for major new research programs in high-performance computing and biofuels.

With each of these tasks completed, UT-Battelle undertook a new challenge, one that in some respects reflected more than any previous project the core goals of the UT-ORNL partnership. For years both parties had discussed the possibility of establishing a new academic program that could capture the potential offered by the combined teaching and research capabilities of the Laboratory and the University. Given the Laboratory's mission, the program would be multidisciplinary in nature and devoted to the study of energy-related sciences. If properly conceived and implemented, the program would

Previous page: UT President John Petersen, Governor Phil Bredesen, Battelle President Jeff Wadsworth, ORNL Director Thom Mason. The genuine partnership among the four institutions was a major factor in ORNL's resurgence.

attract a cadre of the nation's best graduate students, eager to train under internationally renowned researchers at one of the world's most modern and sophisticated scientific facilities. For a university aspiring to be among the country's top 25 public institutions, the program was an opportunity to enrich the University's pool of graduate students and faculty, provide a unique and nationally recognized curriculum, and significantly increase the number of graduate degrees awarded in the sciences. For ORNL, the initiative was a chance to bring to Oak Ridge dozens of extremely bright students who would supplement the Laboratory's research program and, in the case of some, become a talent pipeline for future high-value employees.

In 2010, University and Laboratory officials began serious discussions with Governor Phil Bredesen about the creation of a new UT-ORNL graduate program. The challenges were not trivial. Faculty would have to develop on a fast track the parameters of the program, then design an interdisciplinary curriculum. The parties envisioned a program of extensive interdisciplinary research and coursework with a science and engineering curriculum fashioned around the production, distribution, and consumption of energy. Reflecting the UT-Battelle culture, they conducted interviews with senior-level representatives from industries including Siemens, DuPont, Chevron, GE, and Exxon Mobil to understand what companies would want in a program designed to foster the skills required for public policymaking and entrepreneurship.

Other aspects of the proposed graduate program were equally unconventional. Subject to a process of review and evaluation, ORNL research staff would be granted affiliate faculty status to instruct and mentor students. Students would spend most of their first year attending classes on the Knoxville campus, with the remainder engaged in direct research activities in the laboratories at ORNL. The creative approach enabled significant expansion of graduate faculty and the number of graduate degrees in the sciences with only a marginal increase in university expenditures.

UT-Battelle worked closely with the Department of Energy to ensure the program's process, as well as its intent, was aligned with DOE's expectations at ORNL. In Nashville, formal approval of the new degree by the Tennessee Higher Education Commission, a process that sometimes took years, was compressed into a few months. Most challenging of all, the new graduate program required initial state funding, meaning in this instance passage of legislation authorizing the joint program and the expenditure of some $6 million in state funds. UT President Joe DiPietro and Knoxville Chancellor Jimmy Cheek visited personally with legislators and staff to encourage passage of the legislation.

Governor Bredesen not only supported the program, but assigned his staff to help work out details with the Legislature and UT Board of Trustees. In one of those rare historical coincidences that shape public policy, the Tennessee Governor had earned a degree in physics at Harvard, and thus had a genuine personal interest in overcoming the barriers to establishing a top-flight graduate program supported by the combined resources of the state's flagship university and the nation's largest energy laboratory. With the Governor's assistance, the legislation was enacted and funding provided for the Center for Interdisciplinary Research and Graduate Education, putting in place the foundation for one of the University of Tennessee's most prestigious academic programs.

Following the end of the Governor's eight-year administration, the program was renamed the Bredesen Center, offering doctoral fellowships and highly competitive stipends through two interdisciplinary graduate degrees to some of America's most talented graduate students. The Energy Science and Engineering Ph.D. program engages graduate students in areas of research that include nuclear energy, bioenergy and biofuels, renewable energy, energy conversion and storage, distributed energy and grid management, as well as environmental and climate sciences. Graduate students work directly with teams at both ORNL and UT on problem-oriented research.

The Bredesen Center is also home to the highly competitive Distinguished Graduate Fellowship Ph.D. program, designed to promote leadership in science, technology, engineering, and mathematics. Distinguished Graduate Fellows are heavily involved in research at ORNL while simultaneously taking specialty courses and seminars to complete their Ph.D.s in the fields of nuclear, computational, and materials science and engineering at the University of Tennessee. Partnering with UT's College of Business Administration, the students have the option to pursue an entrepreneurial track to develop skills of value in forming a technology-based company.

Having achieved approval and funding for the program, UT quickly sought to stand the program up as soon as practical. Former ORNL Deputy

Director for Science and Technology Lee Riedinger was selected to head the new program. A football fan, Riedinger launched a nationwide recruiting effort for graduate students with the same intensity as the football team recruited talented running backs. Stellar students from top universities on the West Coast, the Midwest and the Northeast were identified, contacted, and invited to join some of their southern colleagues on the UT and ORNL campuses. They were introduced, not only to some of the world's most advanced research facilities, but also to the Governor's Chairs recipients who would personally participate in their training. The prospective students were offered stipends of $28,000 in addition to tuition and health benefits.

The aggressive recruitment brought results. The inaugural class contained 19 students. Subsequent classes were designed to add 30 new PhD. candidates each year. With each new class, the program gained in depth and stature.

In some respects, the Bredesen Center represented the culmination of the goals expressed some twelve years earlier in UT-Battelle's original proposal for the ORNL management contract. Even in its earliest days, UT-Battelle believed in the potential made possible by a true partnership among the Laboratory, the University, the state of Tennessee and the Department of Energy. With dozens of students and faculty fully integrated in a program of national prominence, the Bredesen Center is evidence that a decade later the foundation has been laid, and that many of those original goals have become a reality.

Right: *The Bredesen Center attracts high-achieving graduate students to UT who want to conduct research at ORNL.*

Corporate Outreach

A VALUED MEMBER OF THE COMMUNITY

❝*We genuinely believed that in order to be a great laboratory, ORNL had to be viewed as a highly-valued member of the community.*❞

—*Dr. Jeff Wadsworth, Director of ORNL, President of Battelle*

One of the most compelling elements of UT-Battelle's proposal to manage Oak Ridge National Laboratory was the emphasis placed on corporate outreach activities in the community and the region. While all proposals contain some mention of corporate outreach, UT-Battelle's plan for ORNL was distinguished by the expression of a philosophy that viewed excellence in community service as a core value equally important as research and operational excellence. The proposal was a reflection of the cultures of both the University of Tennessee and its mission of public service as a land-grant institution, and Battelle, which for a half-century had been a leader in philanthropic activities in the Central Ohio region. Founded on this tradition of public service, UT-Battelle sought to reshape ORNL's mission, not just as an international center of research, but also as the region's foremost leader of corporate outreach.

Not surprisingly, ORNL's new outreach philosophy was heavily influenced by both UT and Battelle. The first change was one of scale. UT-Battelle's initial outreach budgets were in excess of $1.2 million, substantially higher than any previous contractor in Oak Ridge and higher than any corporation in greater Knoxville, particularly as measured by the percentage of UT-Battelle's $7.7 million annual fee that was dedicated to outreach activities. As the state's flagship university, UT naturally preferred that a large portion of outreach activities be earmarked to K-12 and higher education programs. Also a strong advocate for education, Battelle's corporate reputation for technology transfer complemented the University and was an ideal asset

to supplement the state's regional economic development efforts. Battelle added another defining principle. Rather than attempt to fund the dozens of golf tournament and banquet requests that flooded ORNL each year, UT-Battelle instead would focus on what was termed "legacy investments," defined as tangible gifts that would have lasting value.

Enduring Next Year and the Next

The product of this partnership was a corporate outreach program unlike any in the East Tennessee region, distinguished in part by projects that were much larger in size and fewer in number. UT-Battelle's allocations were relatively consistent: roughly 40 percent of the outreach funds were dedicated toward an effort to become Tennessee's premier supporter of science education; a comparable amount was used to fund economic development activities. The remainder was designated for civic and volunteer projects such as the United Way or Habitat for Humanity.

The results of a large and targeted outreach program were felt immediately in the Oak Ridge community. Outstanding students received four-year UT-Battelle scholarships valued at $20,000, with the criteria that one of their parents work at ORNL and that they major in math or science at the University of Tennessee. The University received $150,000 for the Academy of Math and Science, where teachers received summer training to expand their certifications in math and science. Certified science teachers received a $10,000 signing bonus from UT-Battelle if they pledged to teach at least three years in a school that

had gone two or more years without a certified science teacher. Five K-12 schools each year were selected to receive $10,000 to equip a science laboratory. In one rural school that had never employed a certified science teacher, the equipment ordered was too large to go through the doors and too heavy to carry up to the second floor. The school turned the event into a celebration, removing the doors and getting the football team to carry the bulky laboratory equipment up the stairs. Inspired by the event, the community hired the school's first certified science teacher.

Other outreach projects were equally unconventional. Looking at the Oak Ridge community, UT-Battelle determined that targeted investments in the city's outstanding rowing venue offered an immediate opportunity to expand visibility and generate additional revenues for the city. UT-Battelle donated $135,000 to construct a finish line tower, followed by $100,000 to upgrade the starting docks and an additional $130,000 to build a pavilion on the lake. In part as a result of UT-Battelle's investments, Oak Ridge was selected on a recurring basis to host the NCAA rowing championships as well as national high school regattas. The benefits to the city, as measured by additional exposure and tax revenues, exceeded all expectations.

Previous page: *Rowing venue at Melton Hill Lake. UT-Battelle investments helped make Oak Ridge one of America's premier locations for competitive rowing.*

Preserving the Community's Greatest Asset

The success of the legacy investments at the rowing center laid the groundwork for one of the largest and most ambitious civic projects in the city's history. As relative outsiders with fresh eyes, UT-Battelle saw two unique assets that set Oak Ridge apart from many communities. One was a lake that provided some of the smoothest water in America for competitive rowing. The other was Oak Ridge High School, since its creation during World War II one of the outstanding high schools, not just in Tennessee, but nationally, turning out disproportionate numbers of Merit Scholars, Ivy League students and winners of the state Science Bowl competition. UT-Battelle in 2003 viewed the challenges at Oak Ridge High School as remarkably similar to those at the Laboratory. The school still had an outstanding faculty and student body, but the buildings were much like their counterparts at ORNL—constructed as state-of-the-art facilities in the 1950s and looking the part. And just like at the Lab, a city attempting to new attract residents increasingly encountered trouble convincing prospective parents that a high school with such poor infrastructure was worthy of its reputation.

UT-Battelle's perspective was a blend of two concerns. As the city's greatest asset, the school was an indispensable part of the community's economic future that needed to be revitalized. For UT-Battelle, efforts to recruit large numbers of scientists with young children would depend to a great extent on overcoming doubts about the quality of Tennessee's education system. In this sense, the ability to showcase a modern school was important in the same way as providing new buildings at the Laboratory.

Working with the Superintendent of Schools and led by ORNL Director Jeff Wadsworth, UT-Battelle initiated a three-part strategy based upon developing a renovation plan for the school with buy-in from teachers and civic leaders structuring a realistic financial plan to fund the project, and ultimately generating public support for a sales tax referendum needed to provide a sustained revenue stream. UT-Battelle paid $100,000 for an independent assessment of the school's buildings and infrastructure and made available a senior facilities manager to serve as moderator for dozens of discussions about what the faculty and community wanted a renovated school to look like. Working quietly with DOE leadership, Wadsworth met with major contractors in Oak Ridge as part of an ambitious goal of raising $8 million in private donations. Other UT-Battelle staff consulted with state government to take advantage of a program that matched private donations with state bonds. Simultaneously, the pending referendum was approached with the same preparation as a political campaign, with a communication plan, polling, press releases, brochures, and phone banks.

The plan unfolded in 2004. The Oak Ridge School Board adopted a $55 million proposal to demolish and replace a section of the school and renovate the remaining buildings on the high school's existing site. The plan included remodeling the school's auditorium to serve as home to the Oak Ridge symphony, an aspect of the project attractive to many senior citizens. Five and one-half million dollars in pledges from major contractors were combined with more than $3 million in smaller donations from companies and individuals. When matched with the state funds, the amount raised for the renovation project totaled

approximately $17 million, a remarkable sum for a community of fewer than 30,000 residents. The next step was the most controversial, and reflected UT-Battelle's willingness to take a calculated risk with a simple proposition to the citizens of Oak Ridge. The $17 million would be made available if, and only if, voters approved a one-half cent increase in the sales tax dedicated to the high school project.

On election day, with parents making phone calls and high school students serving as poll workers, Oak Ridge voters approved the sales tax increase by a margin of 73 percent. The overwhelming vote in favor of renovating Oak Ridge High School was

a historic statement that expressed support of public education and, perhaps equally important, confidence by the citizens of Oak Ridge in their community's future.

Moving Technologies to the Market

While no other single project approached the size of the high school renovation, UT-Battelle's commitment to promoting the commercialization of new technologies had a similar lasting impact on the region. Over the years ORNL technologies had received more prestigious R&D 100 Awards than any laboratory in the DOE system. Despite this indicator, both UT and Battelle

believed that efforts to transfer ORNL technologies from the laboratory to the market would remain largely ineffective without a new multi-faceted strategy that addressed the specific challenges that had doomed so many start-up companies in the past. In 2000, UT-Battelle designated $400,000 annually for the creation of the Center for Entrepreneurial Growth, designed to provide a support system for researchers who needed assistance with legal, financial, marketing and other issues related to forming a new company. The Laboratory assigned commercial managers to assist with the often complicated process of licensing technologies. To increase needed investment capital, UT-Battelle worked

with East Tennessee business leaders to create the $35 million Innovation Valley Venture Fund. When combined with the affiliate Battelle Ventures, the two funds made available $255 million for the creation of early-stage technology companies.

Borrowing a page from modernization efforts at the Laboratory and the high school, UT-Battelle also sought to create a physical presence for companies wishing to locate near the exceptional talent and facilities at ORNL. Once again using land deeded by the Department of Energy, UT-Battelle developed the nation's only science and technology park that resided within the boundaries of

a national laboratory. With 155,000 sq. ft. of office space, Phase I of the park housed ten high-tech tenants moving a variety of new technologies into the regional economy.

In the first decade, more than 100 start-up companies were licensed and created with the help of UT-Battelle's commercialization program. Despite this impressive number of new companies, officials were confident that efforts to grow new companies and new jobs were just beginning to mature. At the University of Tennessee, the University's Research Foundation

established a new business incubator and embarked upon an ambitious effort to build Cherokee Farm, a new technology park near the campus that would be home to start-up companies located adjacent to the new Joint Institute for Advanced Materials.

At ORNL, researchers were gazing over the horizon toward a new technology that held the potential to change the face of American industry. Working at the new Advanced Materials Training and Education Center, scientists believed that carbon fiber, made from the waste products left over in the

production of biofuels, could prove to be a cost-efficient replacement for steel and aluminum in the automobile and aeronautics industries. Much lighter and with comparable strength, carbon fiber represented an exciting opportunity, not just for UT-Battelle's commercialization program, but for the region's economic future.

Though perhaps not apparent at the time, UT-Battelle's economic development initiatives, like similar initiatives in research and operations, reflected the company's growing willingness to break the mold.

Led by UT-Battelle, the $55 million renovation of Oak Ridge High School was the largest K-12 project in Tennessee history.

The Best Is Yet to Come

> **"***The past is only a prelude. As excited as I am about what UT-Battelle has accomplished, I am even more excited about where they will take ORNL over the next decade.***"**
>
> —*Congressman Chuck Fleischmann*

12

Among the most encouraging comments about Oak Ridge National Laboratory comes from those who visit the campus for the first time since the 1990s. Invariably, they marvel at the dramatic visual contrast with the Laboratory they remember. A manicured lawn, surrounded by four large, modern buildings, occupies the former site of an ugly asphalt parking lot. Down the pedestrian walkway, a new Chemical Sciences building has replaced yet another parking lot. On Chestnut Ridge, the Spallation Neutron Source, the Center for Nanophase Materials Science, and the Joint Institute for Neutron Science—all constructed in the last decade— comprise one of the world's

foremost centers for the study of materials. Down the hill, the sprawling facilities of ORNL have actually begun to resemble a college campus. Miles of fences have come down, replaced by waterfalls, stone walls and greenways surrounding the buildings and bicycles dotting the streets.

The reaction of these visitors is not limited to the visual changes that have taken place. A tour introduces them to a sampling of the magnificent new research programs made possible by the Laboratory's modern facilities. In one building they see the most impressive collection of computational power on the planet. Next door, scientists are using the supercomputers

Previous page: *Cherokee Farm. Located across the river from the University, the new technology park will be home to the UT-ORNL Joint Institute for Materials Sciences.*

to design a new generation of nuclear reactors that will last longer and produce less waste. Across the Quad, one of the nation's premier programs of nuclear non-proliferation extends ORNL's reach into the farthest reaches of Central Asia. One can often find the national security staff in the cafeteria, seated next to researchers who are developing cables that will handle ten times the amount of electrical power, designing car batteries that will go up to 500 miles without a charge, or inventing technologies that make it possible for low-cost houses to produce more electricity than they use.

Perhaps most important of all, visitors to ORNL see the reflection of this tremendous intellectual energy in the employees and students who work at the Laboratory. The anxiety and pessimism that characterized Laboratory staff morale at the turn of the century have been replaced by a pride in the enormous accomplishments that have taken place and a feeling that, despite a host of budgetary and operational challenges that always accompany a large research institution, ORNL is as well-positioned as any laboratory in the country to deliver the Department of Energy's mission and take advantage of opportunities that are waiting just over the hill.

The temptation for UT-Battelle would be to look back over the last decade and be content with consolidating the progress made thus far. This perspective, however, would be at odds with the fundamental philosophy responsible for the creativity and risk that made possible the Laboratory's resurgence. The cultures of both Battelle and the University of Tennessee are grounded in the idea that success comes from constantly pushing the boundaries of innovation. A decision to stand on the laurels of past accomplishments would be counter to this culture and counterproductive to the Laboratory's long-term vitality.

With the wind at its back, UT-Battelle is looking toward the next decade with the sense that ORNL is only now beginning to reach its full potential. Innovations that several years ago were only dreams around a conference table are now within the grasp of researchers in modern buildings, with world-class tools and world-class partners. The notion of building a viable fusion reactor is no longer fantasy. The growing possibility of powering the Laboratory with a Small Modular Reactor would mean essentially eliminating ORNL's generation of greenhouse gases. New technologies for cellulosic biofuels and lightweight batteries will transform the nation's energy sector. The ability to replace steel and aluminum with cost-competitive carbon fiber developed at the Laboratory would revolutionize much of the automobile and aeronautics industries.

These breakthrough technologies are only a sampling of the scientific innovations that lie on the horizon at ORNL. Other—in some ways equally aggressive—initiatives are among UT-Battelle's plans. Each year the partnership with the University of Tennessee edges the University closer to its goal of becoming a Top 25 public university. Against the backdrop of carbon fiber's potential, UT-Battelle's efforts to expand the Laboratory's Science and Technology Park and support a new airport in Oak Ridge could leverage a new era of economic growth in the region.

UT-Battelle looks ahead to this future, guided by the same simple goals that first accompanied them to ORNL in 2000. They aspire to provide the Department of Energy with the world's foremost research laboratory. They will partner with the University of Tennessee to maximize one of the state's most valuable assets. And they will use the extraordinary resources of the Laboratory to foster economic prosperity for the region and the nation.

For the heirs of those UT-Battelle staff who began their journey in the rusted Quonset hut back in the winter of 2000, the best is yet to come.

95

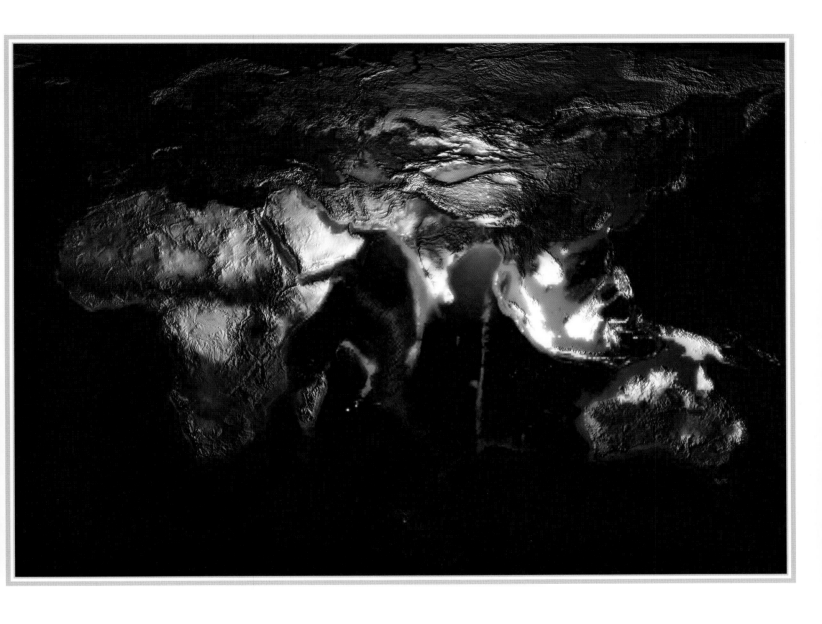